AF567075

Electrohydraulics

First Edition

Electrohydraulics

First Edition

Author: Brian F. P. Andrews

13 Number ISBN: 978-1-901892-20-8

10 Number ISBN 1 901892 20 4

Electrohydraulics

First edition 2008

Other books in association with the BFPA include:

Principles of Hydraulic System Design

Conference Proceedings

Condition Monitoring 1999
Condition Monitoring 2001
Condition Monitoring 2003
Condition Monitoring 2005
Condition Monitoring 2007
COMADEM 2001
QRM 2007

Books in *Coxmoor's* Machine & Systems Condition Monitoring Series include:

Vibration
Wear Debris Analysis
Thermography
Corrosion
Oil Analysis
Load Monitoring
Acoustic Emission & Ultrasonics
Level, Leakage & Flow
Noise & Acoustics
The Concise Encyclopedia of Condition Monitoring

Published by
Coxmoor Publishing Company
UK
Tel: +44 (0) 1451 - 830261
Fax: +44 (0) 1451 - 870661
E-mail: mail@coxmoor.com

Printed in Great Britain by Lightning Source

Foreword

This guideline is one of a series produced by the BFPA Technical committees. These committees are manned by acknowledged experts from industry who give their time voluntarily, as such this document and others form part of a valuable library of technical information. Acknowledgement should also be made to the BFPA member companies whose employees are given time to complete this work.

In addition to being a valuable source of reference to engineers, students may also use this guideline to form part of a study package when completing one of the many BFPA or other awarding body courses.

Ian Morris
Director

Contents

FOUR – OPEN-LOOP SYSTEM CALCULATIONS

FIVE – PERFORMANCE PREDICTION OF CLOSED-LOOP SYSTEMS

SIX – AMPLIFIERS

SEVEN – SENSORS

INTRODUCTION

1. INTRODUCTION

Electrohydraulic control systems have been used in industrial and aerospace applications for more than 40 years. In recent years, however, developments in microelectronics, high power amplifiers, magnetic materials and sensor technology have all combined with advanced manufacturing techniques to change the nature of the control systems marketplace. Competition from electric drives has increased; nevertheless, the same technological advances, which have assisted development of electric drives, have had an impact in the electrohydraulic field enhancing performance and reducing costs. Of particular significance has been the improvement in the reliability of hydraulic components and the practical elimination of leakage through advances in connector design.

The purpose of these guidelines is to give designers and users of fluid power systems an up-to-date appreciation of the performance and application of electrohydraulic control to promote a better awareness of their potential for improving the performance of a wide range of products.

There are numerous existing and potential application areas for which electrohydraulic systems are both suitable and practical. Their increased use has come about, in part, because of the need for accurate control of force and motion to improve process quality and production rates and in part because of the need to integrate with modern electronic controllers.

1.1 Types of Electrohydraulic Systems

Typically, three categories of system configuration can be considered when discussing the application of proportional electrohydraulics but before going further it is important to differentiate between electrically operated switching valves and proportional valves. Switching valves are, by definition, on-off devices, diverting flow from one port to another in response to a digital signal. Proportional valves respond in an analogue manner to a variable input signal, giving flow or pressure in direct proportion to the magnitude of the applied signal. Whilst it is possible to obtain very sophisticated control from switching valves, especially if the switching rate is sufficiently high to give proportional effects in the system being controlled, the application of these valves is outside the scope of these guidelines.

The first category of proportional electrohydraulic system is simply one where manually adjustable pressure and flow control valves are replaced by the corresponding electrohydraulic valves, which allow direct integration with electronic controllers for ease of machine setting and flexibility of operation.

The second category, sometimes described as open-loop control, represents systems where electrohydraulic valves are used in place of switching valves to control the rate of valve opening to achieve smooth acceleration and deceleration. Usually proximity switches are used to detect the position of the moving load and appropriate control actions are then taken by the system controller to decelerate and stop the movement. Fine degrees of control are possible, however, the system is open-loop in the sense that the controller does not know precisely where the load is, only that it has passed a particular proximity switch and been signalled to stop. There is no automatic error detection and correction, consequently variations in load, or friction, or other parameters, such as the update time of the system controller, can affect the precision of control.

The final category consists of what are termed closed-loop systems. Over the years many control strategies have been developed ranging from simple proportional control, through PID control (Proportional, Integral and Derivative) to advanced digital control methods which may be self-setting or adaptive. Despite this, the basic principle is straightforward; the controller sets the demand, that is, the required position, velocity, load or pressure; a sensor or transducer feeds back the current value of the variable; the controller compares the demand with the actual value, determines the error and applies a signal to the control valve, in the appropriate direction, such that the system moves to minimise the error. By actively monitoring where the system is and comparing it with where the controller wants it to be, the loop is closed and errors can be minimised.

1.2 Advantages of Electrohydraulics

Electrohydraulic systems provide the user with many advantages compared with electric drives and electric actuators. The principal advantage is the greater force available for a given actuator size and weight which often allows a more compact system with greater dynamic response particularly where large inertias have to be moved. Properly designed systems, which are well manufactured, will give reliable and leak-free service.

The following characteristics of electrohydraulic systems can often be used to provide performance advantages and lower system costs:

- Higher output force compared to electric drives, often eliminating the need for gear reduction or lead screws and the associated backlash problems.
- Higher output stiffness compared to electric drives giving higher dynamic response in direct drive systems.
- Lower system weight and size, particularly if one flow source serves a number of actuators, or if the actuators can be part of the system structure.
- Simple inclusion of check valves or solenoid-operated valves to maintain actuator position or ensure specific actuator behaviour if electrical power or other prime mover power is lost.
- Ease of heat removal from the point of operation.

In addition, as mentioned above, electrohydraulic proportional valves provide the user with some advantages compared to manually adjusted and electrically switched hydraulic valves. These advantages, described below, can enhance performance and accuracy and offer a system which is significantly easier to use.

- Electrohydraulic valves give smooth operation because the rate of opening and closing of the valve can be controlled by the electronic amplifier.
- Valves can be sited for optimum hydraulic performance without regard to access for manual adjustment making machine design easier.
- Valves can be continuously adjusted during the machine cycle allowing flow or pressure to be better matched to process requirements.
- Accuracy and repeatability of the total system can be improved by the use of electrical feedback transducers measuring position, pressure, force, velocity, flows or any other variable and automatically adjusting the signal to the valve as necessary.

1.3 Guidelines Structure

These guidelines are structured in a logical way, which reflects the physical components which make up an electrohydraulic control system.

Part 1 - describes the types of electrohydraulic valve available on the market.

Part 2 - covers the selection of valves, types of valve actuators and some valve characteristics.

Part 3 - explains the equations used for calculating the velocity of an actuator in a system that uses an electrohydraulic valve. The method of calculating velocity is applicable to both open-loop and closed-loop systems.

Part 4 - explains the performance prediction of closed-loop systems and introduces the concept of the Framework System, a simple method of designing a closed-loop position control system without using complex mathematics.

Part 5 - covers the performance and specification of the drive amplifiers, which act as the interface between the electrohydraulic valve and the control system.

Part 6 - covers, in some detail, sensors and transducers. The choice of these devices is critical to the performance of a control system because in the final analysis, with all other parameters being equal, the accuracy of the system can be no better than that of the transducer.

Part 7 - explains the different types of pressure controls.

Part 8 - gives guidance on the selection of pressure controls.

Part 9 - explains a number of industrial applications of electrohydraulic systems, presented giving practical data of the parameters involved and the performance achieved.

Part 10- is a glossary of terms and is intended to assist a newcomer to this field to come to grips with some of the technical terms employed by hydraulic engineers.

Part 11 - is a recommended Reading List.

ELECTROHYDRAULIC VALVES

1. DESCRIPTION OF VALVES

1.1 Electrohydraulic Control

Electrohydraulic modulating valves are designed to convert an electrical input signal into either a hydraulic flow or a pressure. In both of these cases the hydraulic control is achieved by passing hydraulic fluid through a restricting passage. It is the degree of restriction or, more precisely, the cross-sectional area available to the flow, that gives the modulating control. If a constant flow is passed through a restriction the pressure drop across the orifice will increase as the restricting area is reduced. Alternatively, it can be considered that a smaller flow will result when a fixed pressure is used to push flow through an area which is then reduced. The mathematical relationship between flow and pressure in such a controlling orifice is usually assumed to be a square root relationship. This is true in most cases unless a long flow path or high viscosity fluids are involved. The form of the relationship is:

$Q = k \cdot A \cdot \sqrt{\Delta P}$		where..
Q	=	flow
A	=	area of restriction
ΔP	=	pressure difference
k	=	constant depending on the restriction shape and fluid properties

Equation 1.1

Whether a particular variable orifice is used to control pressure or flow depends on how it is connected within the hydraulic circuit and thus whether a constant pressure or constant flow assumption is relevant. Although these fundamental similarities exist, valves are normally designed to operate for one function only and are not interchangeable. There are, however, some designs which can be used for either flow or pressure control and even some which can be changed between these functions as appropriate.

All of the valves considered here have some means of varying this restricting area by mechanical conversion from the electrical input. All hydraulic valves currently available rely on the use of electromagnetism for this conversion process. The electrical input to the valve is a current which will give an output of force or torque and this in turn is used in a variety of ways to control the valve area. In cases where the output force is small, some further amplification is necessary and this is achieved with an extra stage in the valve. Such multi-stage valves are very common in electrohydraulic systems, particularly when large flows are involved. The operating principles of the most common valve designs are described in this section of the guidelines.

1.2 Proportional Solenoid Devices

A proportional solenoid is a specially modified solenoid which will give an output force which is both proportional to the input current and constant within the working stroke. Like conventional solenoids, they can only produce a force in one direction, electromagnetic attraction, irrespective of the direction of current flow. They are usually used in valves in the sense of pushing to move a part of the valve. However, proportional solenoids can only be produced to work with a DC supply and they are generally used with amplifiers giving a controlled current output. This allows a constant current to be maintained irrespective of temperature and other effects on the solenoid characteristics.

There are two basic methods for using this type of solenoid in a hydraulic valve and these may be distinguished by the terms of force control and stroke control. In the first method the solenoid force is balanced directly, whereas in the second a position feedback loop is used. However, in both these cases the solenoid is still producing a force and the intention is to control the position of a part of the valve. This moving part of the valve may then be used either to control pressure or flow and detailed descriptions follow.

Pressure controls are covered in section 7.

2. ELECTROHYDRAULIC FLOW CONTROL VALVES

2.1 Flow Control

In a flow control valve the armature of a force controlled solenoid acts directly on a compression spring. For the spring the relationship between force and compression is linear and so a given current in the solenoid is converted to a proportional displacement of the armature and anything attached to it. This design relies on the predictable linear

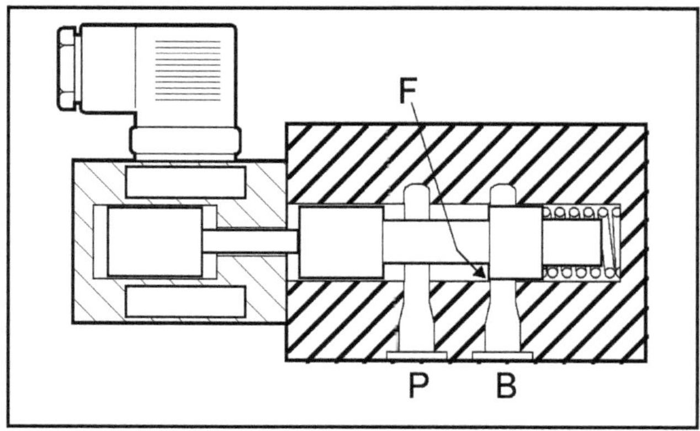

Figure 1.1: Two-port Valve

behaviour for both the solenoid and the spring to give an overall relationship between current and displacement which is linear. As described in the following sections, there are limitations to the characteristics which can be obtained with this design.

The simplest flow control valve is a two-port valve and the basic features are shown in cross-section in figure 1.1. It is called "two-port" because there are two main flow connections to the valve even though additional connections may exist. As can be seen in the figure, if the solenoid current is increased then the moving part of the valve, shown as a cylindrical spool, will be pushed to the right against a compression spring. As the spool moves it will open up a larger area at 'F', between the supply connection, or 'P' port and the delivery connection, or 'B' port. For a fixed pressure drop across the valve any change in area will change the output flow and so the current can be used to give the flow control that is required.

The force available from such a solenoid is sufficient to act directly on the more complex sliding spool of a four-port valve as shown in figure 1.2. This type of spool valve gives control of the direction of flow as well as its magnitude and has four main

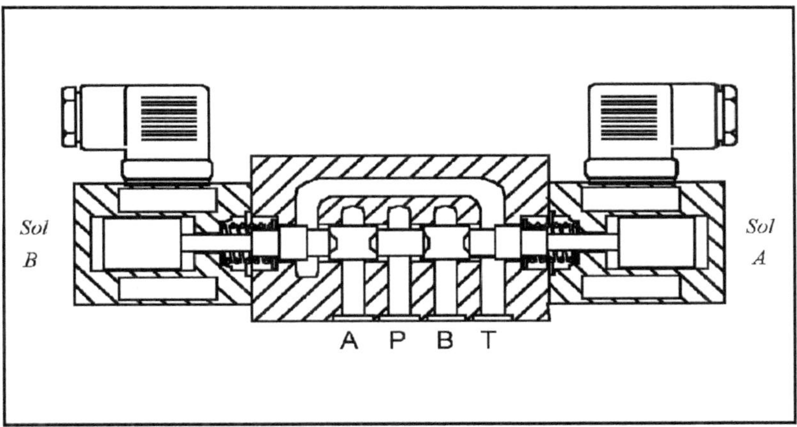

Figure 1.2: Four-port Proportional Solenoid Control Valve

flow connections. As shown, solenoid 'A' is energised pushing the spool to the left against a compression spring and opening up the flow passage between the supply connection, 'P' port and the output connection 'A' port. It also opens the passage between the other output connection 'B' port and the connection to the reservoir, 'T' port, to give a return flow path. As with the two-port valve, the area opened will depend on the spool movement, hence the current supplied will modulate the flow.

Figure 1.2 shows the valve spool in the central position with both solenoids de-energised and all the main flow paths shut. In this condition, it can be seen that the edge of the notches in the spool controlling flow from port 'P' to 'A' does not coincide with the corresponding edge in the valve body. This valve is overlapped. A valve where there is exact coincidence between the spool's metering edges and body is called a zero lapped valve. Alternatively, the valve may be underlapped and in this case there will be a small flow path through all four notches when the spool is in the central position. The overlapped condition as shown in figure 1.2 is lap condition normally used for solenoid-operated valves and requires some spool travel before the flow paths are opened.

The valves in figures 1.1 and 1.2 are known as single-stage valves because there is only one direct stage of amplification within the valve. There is a limit to the flow which can be passed through a single-stage valve and this is determined by the maximum force available from the proportional solenoids. In order to obtain control of larger flows, a two-stage valve is required and a typical example is shown in figure 1.3.

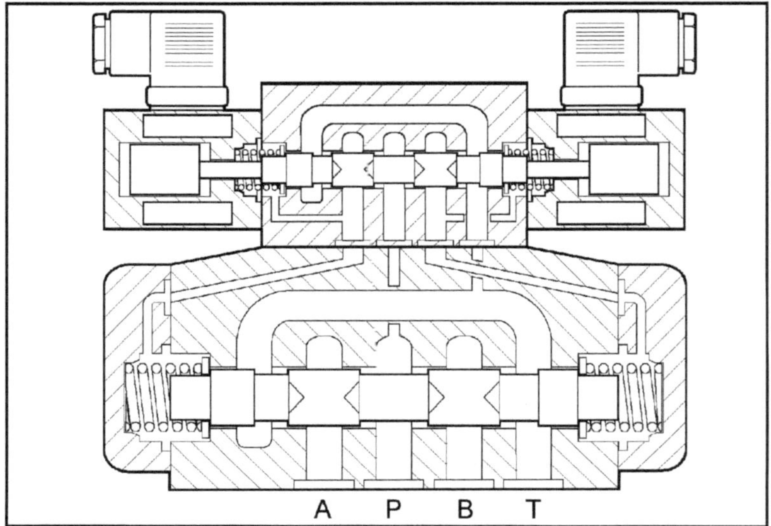

Figure 1.3: Proportional Solenoid 2 Stage Flow Control with Pressure Feedback

The first stage of this valve looks like single-stage valve four-port valve but it is used here to control a flow to the end caps of a second, larger sliding spool. This second spool is another four-port valve and controls the overall output flow of the assembly. It is normal for there to be some interconnection between this spool and the first stage of the valve giving feedback of the second stage spool position. This can be arranged in a number of different ways and the valve in figure 1.3 uses pressure feedback. In this case the second spool moves against compression springs and requires a higher force to give larger displacements and valve openings. This force comes from the pressure acting on the end areas of the second spool itself and can be arranged also to act on the end areas of the first stage spool. In this way it will produce a reaction to oppose the solenoid force and hence give the necessary feedback. A light compression spring only would be used on the first stage in this case.

Many applications use a single rod linear actuator as the output device but this has a different cross-sectional area each side of the piston. The flows at each end of the cylinder will be at the ratio of these areas and in many cases it is preferable to use a valve which has a matching flow characteristic. This is most easily achieved for a 2:1 area ratio because of valve availability. There are also some applications which need only three main flow connections and hence require a three-port valve. Such valves are available but it is more usual to select an appropriate four-port valve, again because of availability.

The alternative stroke controlled proportional solenoid can also be used in single-stage spool valve designs as shown in figure 1.4. In this case there is a displacement transducer built-in to measure the position of the valve spool. This is used as a feedback signal to give closed-loop control of the spool position. It can be seen that the effect of this is similar to the valves already described and the spool opening will again determine the flow. This design will still include a light spring but relies less on the solenoid characteristics for accurate operation. The solenoid will still be driven from a constant current circuit but the current will automatically be adjusted to give the demanded spool position.

This valve can again be used as the first stage of a higher flow, two-stage assembly. Frequently, with a stroke controlled pilot stage, the inter-stage feedback is achieved by use of another position transducer giving a second outer feedback loop to control the second spool position. In all cases where electrical feedback loops are used, all the signal conditioning and control circuitry are built into the amplifier supplied to drive the valve. Interfacing to this type of drive amplifier is usually a simple matter of providing a correctly scaled voltage signal.

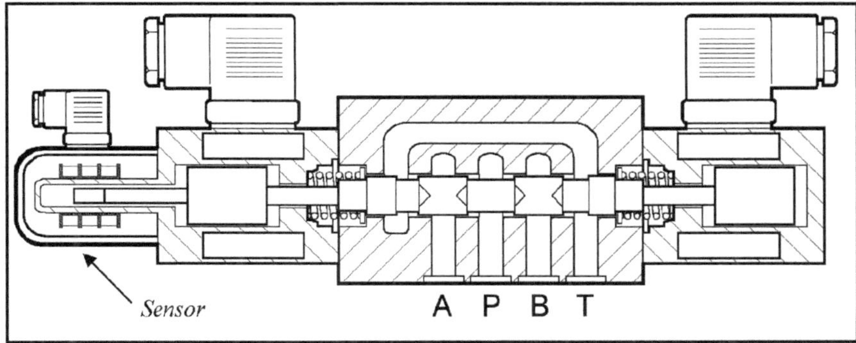

Figure 1.4: Four-port Proportional Stroke Controlled Solenoid Valve

2.2 Torque Motor Devices

The majority of valves built with torque motor inputs are used for flow control. However, they can be used for pressure control and other special versions are available. The torque motors used in hydraulic valves are generally small, low power devices, capable of rotating through small angles and usually used in designs with a high hydraulic amplification to achieve a useful output. However, the small physical size does allow torque motor valves to be used where space is a critical consideration. A typical torque

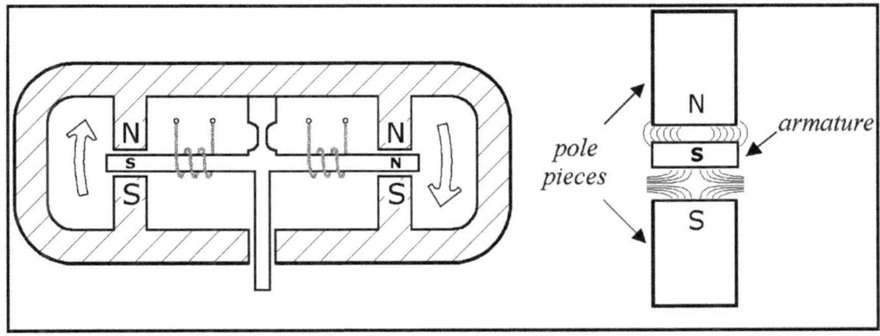

Figure 1.5: Torque Motor

motor construction is shown in figure 1.5. It comprises of a pivoted armature, with two wound coils and the ends of the armature situated between the pole pieces of a permanent magnet. The output is produced as a result of the interaction between the armature, acting as an electromagnet and the field produced by the permanent magnet. Like the solenoid, a torque motor requires a DC input but unlike the solenoid it can produce a torque in both senses of rotation depending on the direction of current flow in the coils.

2.2.1 Nozzle Flapper Valve

The double nozzle flapper configuration is the most common hydraulic element used with torque motors and is shown as an assembly in figure 1.6. The flapper is rigidly attached to the armature of the torque motor and moves between two small diameter nozzles. The nozzles are supplied with high pressure hydraulic fluid and they discharge into a low-pressure chamber connected to the system reservoir. When the flapper is moved by the torque motor it will reduce the flow area through one of the nozzles and increase the area from the other.

When used with a pair of fixed orifices in the supply to the nozzles, as shown in figure 1.6, the pressure upstream of the reduced nozzle will increase and conversely, the pressure upstream of the other nozzle will reduce. The pressure forces produced on the flapper will tend to oppose the electromagnetic torque and the flapper deflection will reduce. The pressure difference produced is proportional to the input current and can be used either to supply flow to a small actuator or as a pilot pressure signal. Additionally, it is possible to use a spring-loaded flapper with a single nozzle for either flow or pressure control.

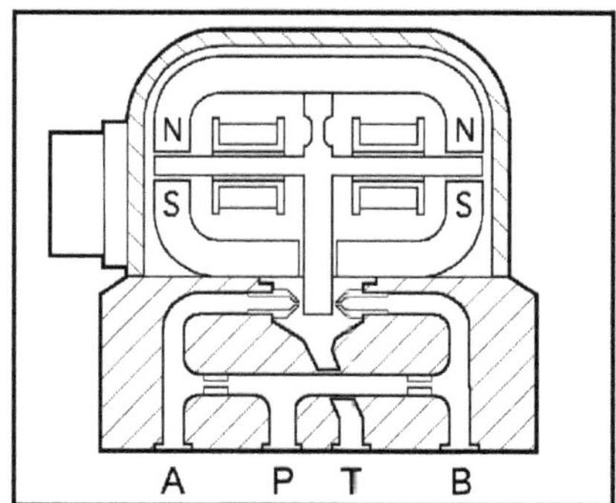

Figure 1.6: Double Nozzle Flapper Stage

2.2.2 Nozzle Flapper Two-Stage Valve

The double nozzle flapper can be used to control a second stage, four-port, sliding spool valve and a typical arrangement is shown in figure 1.7. The end caps of the spool valve are connected upstream of the nozzles so that the pressure difference produced by the first stage acts to move the spool. The spool and the flapper are physically connected by a carefully shaped flexible link sometimes called a feedback wire. When the spool moves, this link transmits a torque back to the first stage to centre the flapper giving an equilibrium condition when the electromagnetic and feedback torques balance. This has the effect of direct mechanical position feedback between the spool and

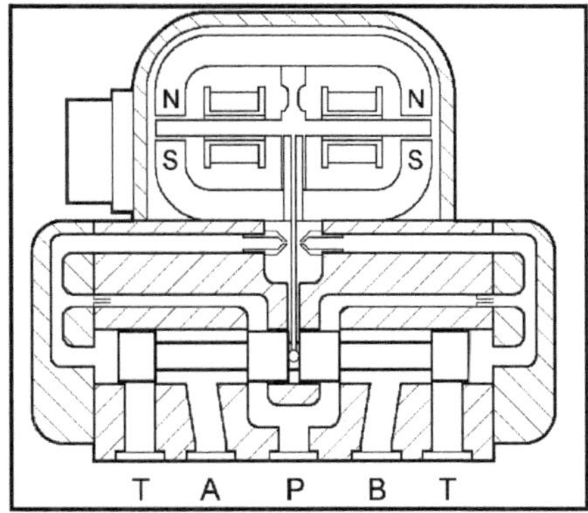

Figure 1.7: Nozzle Flapper Two-Stage Valve

the double nozzle flapper first stage. The input current to the torque motor thus produces a proportional displacement of the spool and opens up the flow area giving proportional flow for a fixed pressure difference across the valve. The spool lap in these valves is very accurately controlled and zero lapped valves are not uncommon.

The type of feedback shown here is also used in some designs of proportional solenoid valves and the alternatives which are used also have similarities. The spool position can be measured with a transducer and closed-loop electrical control used. The spool can be spring centred so that the feedback is achieved by the pressure difference across the end caps of the spool acting back to the flapper. In both these cases, the valve is used primarily for flow control. In cases where a required load pressure is to be controlled it is also possible to use this pressure as the feedback signal to both the spool and the flapper.

2.2.3 Jet Pipe Two-Stage Valve

There are a number of servo-valves manufactured with different first stage designs but still using a torque motor. Most of these are similar in principle to the nozzle flapper, however, the jet pipe first stage is significantly different. A jet pipe servo-valve is shown in figure 1.8 and at first sight the use of torque motor and spool make this valve very similar to the nozzle flapper valves. In this case, however, the end caps of the spool

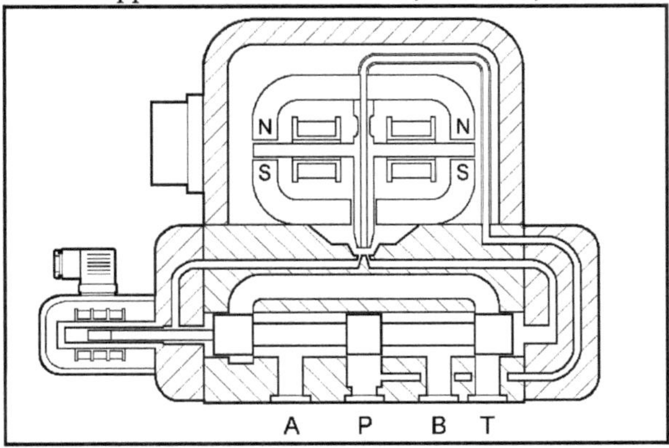

Figure 1.8: Jet Pipe Two-Stage Valve

have only one connection each and are supplied by the single moving nozzle attached to the torque motor armature. This nozzle, or jet pipe, is supplied with high pressure fluid which leaves the nozzle with high kinetic energy. When the nozzle is deflected, this high velocity jet of fluid is directed preferentially to one end cap. The pressure recovery as the fluid is brought to rest in the dead volume of the end cap is sufficient to move the spool. The position of the spool is measured by a position sensor shown on the left-hand end of the body. In this instance the position sensor is an LVDT as described in section 6.2.2. The signal from the LVDT is used in an electrical closed-loop which moves the nozzle back to a central position when the spool reaches the desired position. When the nozzle is centralised the jet is equally distributed to both end caps and they will both be at the same pressure.

2.3 Three-Stage Valves

When it is necessary to control large flows in high power systems, then a further stage of hydraulic amplification can be used. Another large four-port sliding spool valve is added to the assembly. This valve controls flow in the manner described and in turn its position is determined by a supply to its end caps from the previous stage. Again some form of interstage feedback is necessary. This is usually either electrical or pressure feedback.

2.4 Load Independent Flow Control Valves

As indicated in the first paragraph of this section, flow through a hydraulic valve is dependent on both the area available and the pressure difference across the valve. Most of the above descriptions for flow control valves have considered both a constant supply pressure and a negligible or constant load pressure. In general this will not be true and for a given valve opening the actual flow will vary with any changes in these pressures. However, there are electrohydraulic valve designs which compensate for any changes in pressure drop across the valve and will control to a fixed output flow for a particular demand. Two main methods are used:

Firstly and more commonly, this independence is achieved by maintaining a constant pressure drop across the critical flow area irrespective of system pressure changes. The modified valve assembly is shown in figure 1.9 and contains an extra two-port spool valve, sometimes called a hydrostat. This compensating spool has a light spring pushing it to the shut position and when open connects the valve supply directly back to the system reservoir. One end of this spool is open to the supply pressure and the other to the load pressure. These two pressures are also connected across the main flow control spool and establish the pressure difference across the metering area. The compensating spool is controlled by this electrohydraulic pressure difference and effectively sets the valve supply pressure to a value just above the load pressure. The difference between these pressures is determined by the preload spring and is usually of the order of 3 to 12 bar. Since this pressure difference is now constant, established by the compensating spool, the flow from the main spool becomes dependent only on its opening, which should be established by the valve demand. This type of compensation is called "by pass" compensation and is a very energy efficient valve control technique. However, it is more difficult to use where more than one valve is supplied from a single

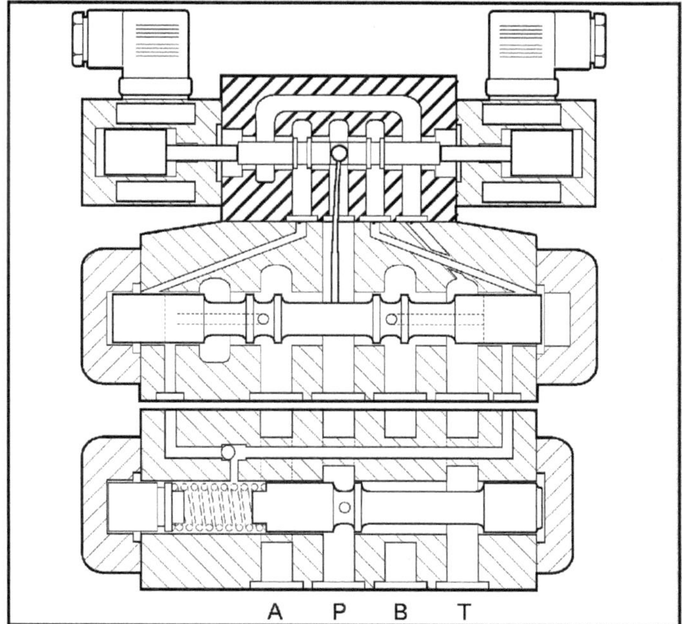

Figure 1.9: By Pass Pressure Compensated Valve

pump. There is an alternative called "series" compensation which can be used with multi-valve systems and some manufacturers incorporate both features in a single valve block.

The second method of achieving load independent flow control uses a flow sensor to give a feedback signal. The opening of the controlling orifice is then adjusted to maintain a constant flow by a closed-loop controller. Although this can be achieved electrically with separate components, there are valves with flow sensing built-in and these operate as a mechanical closed-loop system.

SELECTION OF VALVES

1. INTRODUCTION

This section is concerned with the selection of four-way electrohydraulic valves which include valves described as servo-valves and as proportional valves.

While it is possible to differentiate between servo-valves and proportional valves based on the mechanical construction as described in section 1 of this guideline, from a performance viewpoint there is not a clear distinction between these valves. Originally proportional valves were developed from solenoid actuated directional valves and consequently tended to have a large overlap, a non-linear characteristic and a poor dynamic response compared to servo-valves. Whilst some proportional valves still have these properties, many of them have improved to the extent that given a particular application it is often possible to find a proportional valve that provides an equivalent performance to the servo-valve.

As a general rule servo-valves can have a higher frequency response but the construction makes them more susceptible to contamination particles in the hydraulic fluid. For the same dynamic performance a servo-valve may have a larger quiescent flow.

2. VALVE CHARACTERISTICS AND TEST METHODS

The method of testing the performance of proportional and servo-valves is described in ISO 10770-1 for four-way valves and ISO 10770-2 for three-way valves.

The performance parameters can broadly be divided into flow characteristics; dynamic characteristics; null shifts and hysteresis; linearity, resolution and threshold; proof pressure and electrical characteristics.

2.1 Flow Characteristics

The flow through an electrohydraulic valve is not just related to spool position it is also a function of a number of external factors including; supply pressure, load pressure drop and actuator area ratio.

The most widely quoted flow characteristic for an electrohydraulic valve is the maximum output flow at the rated valve pressure drop. This is known as the rated flow and is measured with the A and B ports subject to the same flow. The rated pressure drop is selected by the manufacturer and traditionally 70 bar is used for servo-valves while proportional valves use 10 bar.

Early proportional valves were intended for use in systems with low dynamic response. Consequently the valves were able to give satisfactory control of the load with less pressure drop across the valve. Additionally, systems using proportional valves were often only required to control actuator velocity rather than position so simplifying the control task. If accurate control of actuator velocity was required this could be obtained by using a pressure compensated control flow in conjunction with the proportional valve. The result of the above was that 10 bar valve pressure drop was sufficient for most applications. In some systems this lower valve pressure drop provided some efficiency advantages. However, high performance proportional valves that are intended for use in position control systems much like traditional servo-valves are usually designed to give a 70 bar valve pressure drop at rated flow.

It should be noted that the rated pressure drop is not related to the maximum supply pressure which is normally based on the integrity of the valve body.

As stated above the flow through an electrohydraulic directional valve is related to spool position. However, we do not have direct control over the spool and consequently it is usual to measure flow as a function of input command signal. Figure 2.1 shows the typical variation of valve flow with input signal assuming the two control ports A and B

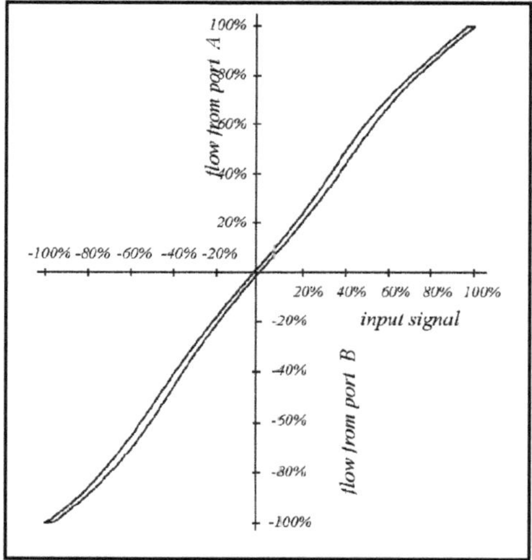

Figure 2.1: Flow versus input signal

are subject to the same flow. The flow rate will often show an approximately linear relationship between signal and flow but this can be distorted by saturation in the valve body or special spool types. For example, in some applications it may be useful to have a lower flow gain over the first 10 or 20%. This can assist with accurate positioning of a cylinder or control of pressure.

It should be noted that the relationship between flow and the system variables listed above is modified by the actuator area ratio and the lap condition. However, any discussion on how area ratio and lap condition effect flow is left until later in this section.

The flow versus input signal characteristic shown in figure 2.1 passes through the origin of the graph indicating the flow reverses exactly when the input signal reverses.

This characteristic is usually associated with a zero lapped spool. A zero lapped spool is usually the best choice if we require a highly accurate actuator positioning system.

2.2 Pressure Gain

To drive the flow and the actuator in opposite directions each side of zero input signal requires a good pressure gain. The pressure gain is defined as the change in input signal required to produce a change in pressure difference between the A and B ports of ±80% of the supply pressure. Between ±100% input signal the pressure difference between the A and B ports will change from +100% of supply pressure to –100% of supply pressure. For a valve with a good pressure gain the bulk of this pressure reversal will occur over a small range of input signal, typically 4% or less and ideally the pressure difference should be zero when the input signal is zero.

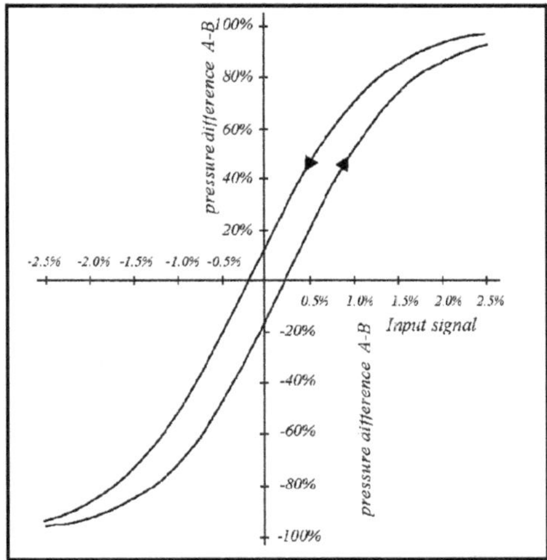

Figure 2.2: Typical pressure gain shown as percent of supply pressure

Pressure gain is measured with the A and B ports connected to small closed volumes that are isolated from each other.

2.3 Lap Condition

In some applications it may be necessary to ensure that the actuator does not creep when the input signal is zero or close to zero. This requirement is met if the spool is overlapped rather than a zero lapped spool. The characteristic of a typical overlapped spool is shown in figure 2.3.

An overlap between the spool and the body of the valve results in a dead band around the null of the valve. The lap can be up to 20% or 30% on low cost valves as this allows the spool and body to be produced with larger tolerances. Using more accurate machining overlaps as small as 2% can be produced if required.

In some instances it may be useful to specify a version of an overlapped spool that drains A and B to tank while the spool is in the overlapped region. This need only be a small flow part but can help minimise the possibility of actuator movement when zero signal is applied or when electrical power is lost. Alternatively, this arrangement can

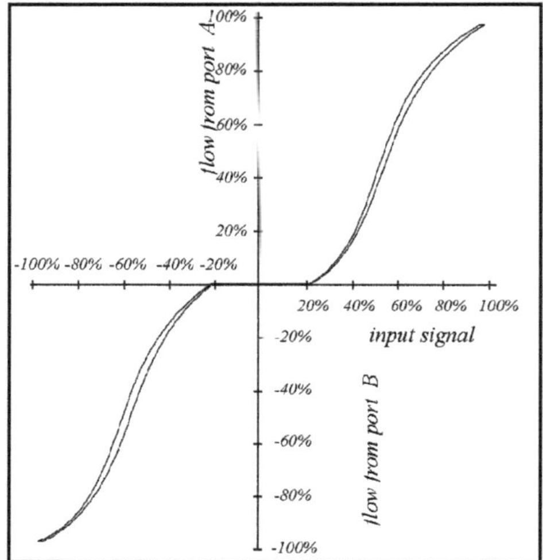

Figure 2.3: Flow characteristic of an overlapped spool

allow the actuator to move slowly under any external load. Sometimes a biased lap condition such as a small connection from A to pressure and B to tank may be useful to ensure the actuator moves slowly in one specific direction when the valve is at null.

Testing the valve with A and B interconnected as shown in figure 2.2 will not show whether A or B is connected to tank within the overlap zone. If required such characteristics can be specified by a number of graphs showing the valve flow through each individual flow part.

2.4 Effect of System Pressure

The valve flow also increases if the system pressure increases. Increasing system pressure increases the pressure drop across the valve and gives the characteristic shown in figure 2.4.

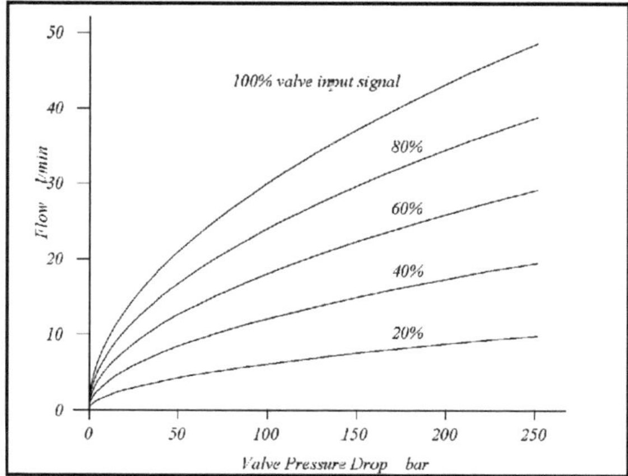

Figure 2.4: The variation of control flow with supply pressure for a valve having a flow of 25 l/min at 70 bar

This relationship usually equates to a square root law. In some cases the flow may deviate significantly from this square law relationship particularly if the spool is able to move under the influence of Bernoulli forces. This can be particularly noticeable in a single-stage valve where the solenoid acts directly on the spool. Consequently when designing a system reference should always be made to the measured or catalogue values of flow rather than assuming the square root relationship holds for all pressures. ISO 10770 recommends testing the flow characteristic as shown in figure 2.4. The flow is measured with the A and B ports interconnected. Ideally two separate differential pressures will be measured during the test corresponding to the two separate flow paths through the valve. Figure 2.4 will be repeated for negative input signals and negative flows.

2.5 Load Pressure Drop

An alternative method of presenting the data shown in figure 2.4 is to assume the system pressure is constant and the pressure drop across the actuator varies. The resulting graph is shown in figure 2.5. The square root relationship is a reasonable approximation but again there can be significant deviations from this.

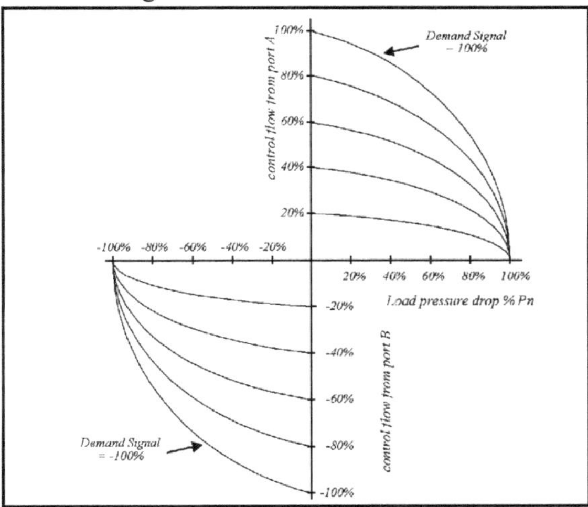

Figure 2.5: Control flow versus load pressure drop

2.6 Quiescent or Leakage Flow

The construction of a servo-valve dictates that even when the valve is stationary there is a control flow through the nozzle/flapper which equates to a constant leakage to tank. There is, for all valves, some leakage across the lands of the spool, particularly if the spool is zero lapped or underlapped. These two components are usually added together and termed internal leakage or quiescent leakage. The leakage across the spool lands is usually largest when the valve is in its central (i.e. null) position. If only one value of leakage is quoted it may be assumed to be the leakage at null.

The leakage is usually small compared to the nominal flow of the valve itself, typically in the range of 0.5 to 3 L/min, although this upper limit may be exceeded for high performance valves. It should be noted that the null leakage will vary with the square root of supply pressures. If the exact leakage is important and the nozzle/flapper and spool are subject to different supply pressures the two leakages will have to be calculated separately.

In a proportional valve there is no nozzle/flapper and hence, given equivalent spool lap conditions, the leakage should be less. This means that single-stage proportional valves with an overlapped spool should have a very low leakage. The reduction in quiescent flow can equate to a significant energy saving if the system is approximately static during a high proportion of its cycle.

Two-stage proportional valves as shown in figure 1.3 are available for a wide range of rated flows, mainly between 40 L/min and 800 L/min. The pilot can be a solenoid actuated spool valve or a servo-valve of the type shown in figure 1.7. When a servo-valve is used the valve is often referred to a three-stage valve.

Additional pilot flow is required when any of the spools in the valve are shifted. The magnitude of the flow can be calculated directly from the velocity of the spool and the area of the end of the spool.

When selecting a two or three-stage valve be aware that the pilot pressure may have a direct impact on the maximum main stage flow. Insufficient pressure may allow the main stage spool to malfunction if the Bernoulli forces exceed the force available to shift the spool. The manufacturer may specify malfunction flow rate as a function of pilot pressure.

The maximum-rated pressure does vary from valve to valve. A 350 bar maximum is now quite common, however, mechanical construction details may limit the maximum supply pressure to the pilot and the maximum pilot return pressure.

2.7 Dynamic Performance

ISO 10770–1 and ISO 10770–2 defines two methods for measuring dynamic response - frequency response and transient response methods. The manufacturer may give frequency response and/or transient response data. Either can be used to predict system performance.

The frequency response is the relationship between the input signal and the spool position over a defined frequency range. If spool position is not available the response can be measured using an actuator to measure the flow between the A and B ports. The actuator must have low inertia and low friction so that the influence of the actuator can be neglected. To define the frequency response it is usual to quote frequency at which the spool position lags 90 degrees behind the input signal.

The frequency response is affected by the peak to peak valve amplitude during test. ISO 10770 allows for measurements from ± 5% peak to peak to ± 75% peak to peak of output rate. The frequency response at 25% will normally be higher than the frequency response at 75% often by a factor of two, depending on valve design. If the manufacturer only quotes one frequency response it would be prudent to assume that this is the response at 25%, hence the response at 75% will be about half of the quoted figure.

The frequency response of a two-stage proportional valve or servo-valve, as shown in figures 1.3 and 1.7 respectively, will normally increase with increasing pressure and conversely degrades with decreasing pressure unless additional valves are added to control the pilot supply.

It is common to quote the response at a supply pressure of 70 or 100 bar but lower pressures can assist with system efficiency.

The range of available proportional and servo-valves is so great that it is not possible to relate frequency response to rated flow. However, figure 2.4 gives an indication of the upper limit for generally available products. Valves able to operate above the shaded area will normally have been made for a specific application such as automotive crash testing.

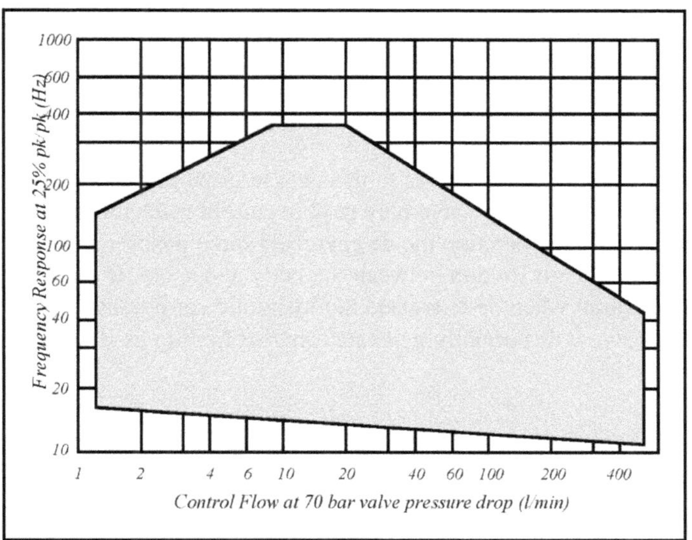

Figure 2.6: Frequency Response of Generally Available Electrohydraulic Valves

Figure 2.6 can be converted to step response time using the equation:

step response time (secs) = 0.6 / Frequency Response (Hz)

Finally it should be remembered that a high frequency response is not the first objective for many applications. A slower valve may provide advantages such as more accurate spool position control or lower quiescent flow.

2.8 Null Shifts and Hysteresis

For zero lapped valves zero command signal should always equate to zero flow. If something in the valve changes it is possible that the zero command signal no longer equates to zero flow this is known as a null shift. If the valve is used in a closed-loop position system any null shift will be seen by the user as a change in actuator position. Null shifts tend to be associated with servo-valves that have an internal mechanical feedback system to establish the spool position. Valves with electrical position sensors tend have less shift.

Null shifts can often occur as a result of changes in; fluid temperature, supply pressure or return line pressure. These factors can change abruptly producing a dynamic error in the system. The magnitude of this error may be reduced by external feedback but a small steady state error may remain.

For valves with overlap the null shift is of no consequence unless it moves the valve outside the overlap region.

Typical null shift values for a servo-valve are quoted below in terms of a percentage of rated input current. A null shift of 2% can be induced by the following:

- supply pressure change of ±20% of nominal
- return line pressure change from 0 to 20% of nominal supply
- temperature change of 30 degrees C

The majority of valves are designed to give a known hydraulic configuration if all electrical power is removed. This de-energised position may or may not be the same as the energised position with a zero command signal. In some applications it may be necessary to set the de-energised position to a non-zero flow condition so that the actuator moves to a safe position.

In most valves the spool moves to this de-energised position under the action of springs and consequently the valve may require careful mechanical adjustment to set the desired position. In operation the de-energised spool position may include a small random variation due to friction between the body and spool. If the valve goes to an offset spool position when de-energised the hydraulic configuration may be well-defined but the valve will normally generate a hydraulic blip as it moves to the offset position.

2.9 Hysteresis

To measure the maximum hysteresis a valve will normally be slowly cycled over the full range of input signal. The hysteresis is measured at any point in the cycle and equals the difference between the increasing signal and the decreasing signal at one particular spool position. This position is often null but any position can be used.

In figure 2.1. the hysteresis is the horizontal distance between the increasing trace and the decreasing trace. Hysteresis has the same effect as a null shift and can introduce actuator position error proportional to the magnitude of the hysteresis. Typical maximum values are 3% to 6% for valves without electrical feedback. Often hysteresis can be reduced by the use of dither (see section 2.3.9).

Valves that have electrical spool position feedback can have hysteresis in the range from 0.1% to 1%.

2.10 Linearity, Resolution and Threshold

The linearity of the valve is usually specified by a plot of the control flow vs. input signal as shown in figure 2.7. Above an input signal of about 10% the variation in the slope of the flow/signal characteristic is, in most cases, fairly small, perhaps around ±10% but this could widen at large flow rates. At small input signals the slope of the flow characteristic can vary as a consequence of the tolerances in the body and spool that contribute to a variable lap condition. Additional underlap tends to increase the flow gain close to null, while overlap tends to reduce the flow gain.

Variations in the lap also affect the pressure gain of the valve. An underlap will reduce the pressure gain and hence allow the actuator position error to increase. Conversely, overlap tends to increase the pressure gain but will eventually result in different pressures for increasing and decreasing signals and this will add to the apparent hysteresis.

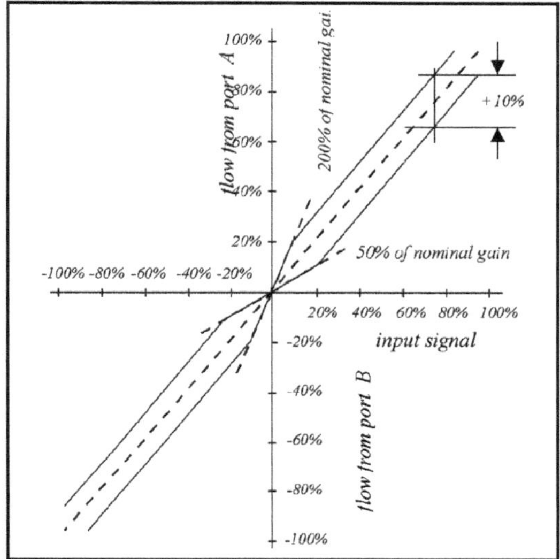

Figure 2.7: No-load flow gain variation

ISO 10770 describes testing for resolution and threshold. The former is the minimum increase in input signal required to produce the smallest perceptible increase in output following a previous increase in output. The threshold is the minimum change in input signal required to produce the smallest perceptible increase in output following a previous decrease in output. As resolution applies to uni-directional movement it is usually smaller than the threshold that will normally include the hysteresis associated with a small reversals. Typical values of threshold are in the range of 0.2% to 2%.

Dither can be used to reduce both resolution and threshold but should be used with care as described in section 2.3.9.

2.11 Proof Pressure, Endurance, Impulse and Environmental Tests

Proof pressure tests are concerned with the valve's ability to operate without external leakage or permanent deformation. ISO 10770 recommends a proof test is conducted with a pressure of 1.3 times the maximum-rated pressure applied to the supply and control ports and a lower pressure to the return ports.

Fatigue tests are concerned with operating the valve for a large number of cycles. This type of test, often run for 10 million cycles, is used to establish the integrity of the valve body.

Fatigue tests are not included in ISO 10770 but are the subject of standard number ISO 10771.

2.12 Electrical characteristics

The electrical parameters covered by ISO 10770 include coil resistance, coil inductance and insulation resistance. Many valves have a number of options for coil resistance with different maximum currents. Normally lower resistance equates to lower inductance and both assist in reducing response time. Conversely, higher resistance coils usually require higher drive voltages and give longer response times.

3. THE CHOICE OF THE VALVE (SYSTEM CONSIDERATIONS)

In this section we will focus on the selection of four-way electrohydraulic valves. This type of valve will normally be used when bi-directional actuator movement is required. If actuator motion is uni-directional then the same style of valve can be used to start and stop the actuator and provide electrical speed control. If small reversals are allowable and required to obtain a specific actuator position then a four-way valve will be used. If reversals are not allowed then the same valve can
have an alternative spool configuration that, for example, allows flow P-A and B-T but not P-B or A-T. Such valves are often referred to as throttles and still allow the actuator to be connected between the A and B ports. Consequently much of the content of this guideline is applicable to this type of throttle valve.

It is also possible for a four-way flow control valve to control both flow and pressure. In an application these may equate velocity and force. This function will usually be achieved with a special spool in a four-way valve and such valves are typically referred to as P/Q valves. In this type of application pressure control is achieved using electrical feedback from a pressure transducer. It is not possible for any valve to control both flow and pressure independently at the same time. If, for example, flow is controlled the pressure is determined by the resistance in the system.

To select a valve for a particular application we need to consider both the valve type and the size of valve required. The first step is to produce a reasonably comprehensive specification for the system requirements. In this context, however, it is important not to build in excessive safety margins to the specification which could, for example, result in a larger or higher capacity valve than necessary being specified which can give a lower system performance.

In pressure control systems an advantage of a single-stage solenoid-operated valve is that the spool can be pushed open by the solenoid even if there is no system pressure. This allows the system pressure to be reduced to a very low level.

A four-way control valve can be used with or without a feedback from, for example, an actuator position sensor. Without any such external feedback the system is said to be open-loop. The selection requirements for open and closed-loop systems need not necessarily be the same. Where this is the case they are discussed separately.

3.1 Type of Actuator

3.1.1 Single Rod Actuators

In the majority of applications, cost and space considerations lead to the use of a single rod actuator. These actuators can be used with a number of different valve arrangements. It should be noted that if this is used with a standard, i.e. symmetrical, flow control valve there may be some reduction in maximum achievable performance.

3.1.2 Single Rod Actuators with Symmetrical Valves

In this configuration, the valve is connected to both sides of the actuator, figure 2.8. This can be used effectively for controlling velocity in either direction but for the same

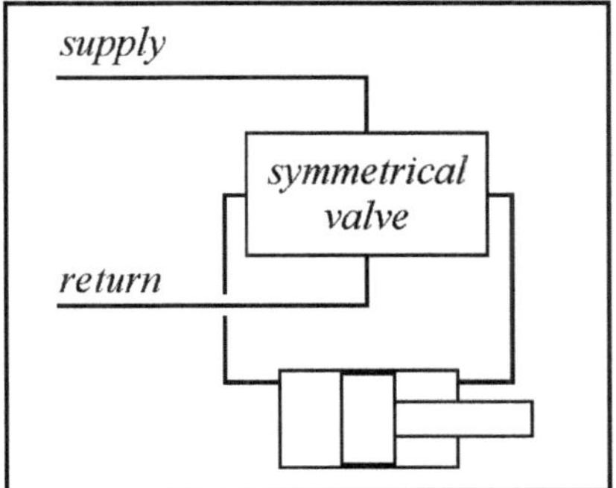

Figure 2.8: Symmetrical valve used with single rod actuator

valve opening these velocities will not be the same. There can be a large change in actuator pressures as the actuator changes direction which can affect the performance if used for position control systems.

Nevertheless, the performance is adequate for many applications and this arrangement is widely used in view of its lower cost.

3.1.3 Three-Port and Regenerative Systems

The alternative method of using a valve with a single rod actuator is in three-way mode as shown in figure 2.9. Here the annular side of the actuator is permanently connected to the supply pressure and the cap end of the actuator is connected to one of the valve's

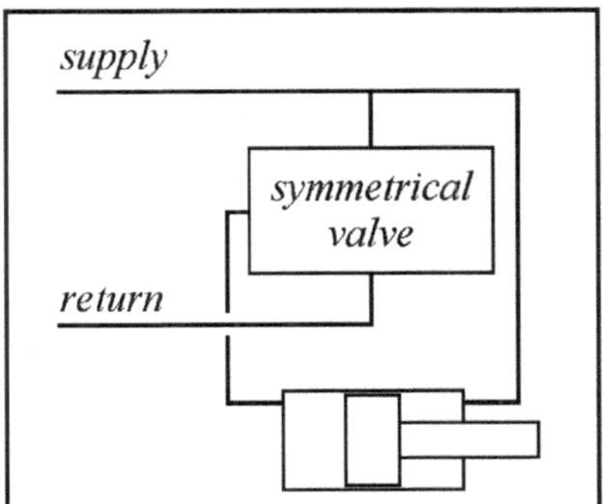

Figure 2.9: Only 1 output of a valve being used to operate a single rod actuator in a regenerative mode

control ports e.g. port A. The valve provides either flow P-A or A-T. The actuator sizes are chosen so that the average full end pressure is approximately half that of the supply pressure.

This method does not have a large change in pressure with change in direction but requires twice the area for the same force capability on extension than in 3.1.2 above. The potential dynamic performance may also be affected by the lower stiffness of the oil columns. When the cylinder is extending, regenerative operation is obtained, flow out of the rod end is circulated back to the supply side of the control valve to supplement the pump flow.

3.1.4 Single Rod Actuator with Asymmetrical Valve

Where high performance is required then arrangements other than those of 3.1.2 and 2.3.1.3 could be more appropriate. One of these is the use of a single rod actuator with an asymmetrical valve which has smaller area ports and hence has lower flow rate through the valve ports connected to the annular end of the actuator, figure 2.10.

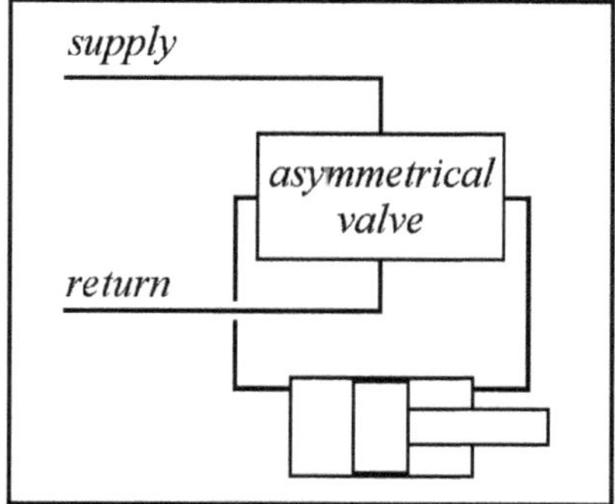

Figure 2.10: Asymmetrical valve used with single rod actuator. The flow ratio of the 2 outputs from the valve is chosen to be equal to the actuator area ratio

To maximise the advantage of this arrangement the ratio of the valve flow rates should be close to the actuator area ratio. Valves of this type are available as standard from some manufacturers and on special order from others. Optimum response occurs with the average actuator pressures equal to half the supply pressure. This arrangement is particularly suitable for applications where a large uni-directional steady output force is required from the actuator. Another form of asymmetrical valve can be used to reduce the prospect of cavitation in applications with large overrunning loads.

3.1.5 Double Rod Actuators

Another arrangement used to achieve high performance is a double rod actuator with a symmetrical valve, figure 2.11. As the actuator has equal areas the total system has the

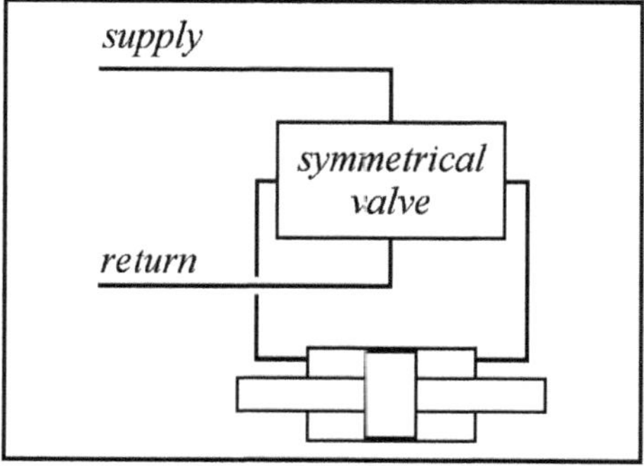

Figure 2.11: Symmetrical valve used with an equal area double rod actuator

advantage of being symmetrical which tends to minimise the effect of some non-linearities and also maximise the dynamic response. However, the actuator does require more space.

3.1.6 Motors and Rotary Actuators

A wide range of types and sizes of motors are available. Different types of construction are used for different speed ranges. These vary from lower than 1 rpm up to more than 10,000 rpm. The use of low speed motors can eliminate the need for reduction gear boxes. Motors can rotate continuously whilst rotary actuators have a limited angle of rotation, usually less than one revolution.

Both motors and rotary actuators are, for selection purposes, equivalent to double rod symmetrical actuators and a symmetrical valve would normally be used.

3.2 Supply Pressure

There are a number of factors that should be considered in the choice of supply pressure, not the least being the previous practice in a particular sector of industry. If an existing power supply is to be used not only should the available flow and pressure be sufficient for the application but the level of filtration and contamination control should be suitable for the intended electrohydraulic system (see BFPA/P5 'Guidelines to Contamination Control in Hydraulic Fluid Power Systems'). A separate power supply may be a cost-effective alternative to cleaning up a very large system.

In aircraft and defence applications where size and weight are at a premium, supply pressures are typically in the range 200 to 270 bar. At the other end of the spectrum where high output stiffness is required, e.g. some machine tool systems, output actuator areas (or displacements for rotary devices) are made large and hence pressures may be as low as 20 bar. For general industrial use, lowest cost is usually obtained with a supply pressure around 100 to 200 bar.

The supply pressure should be sufficient to provide both valve pressure and load pressure drop.

3.3 Actuator Size

The actuator size must be selected to provide the required output force for the allowable load pressure drop. In high performance systems, excess actuator sizes are sometimes used to increase the stiffness of the trapped oil volume and hence improve the dynamic response.

If the actuator is selected from a limited range of sizes the resulting actuator may be larger than ideal. Unfortunately, a larger actuator area requires a larger valve flow to obtain the same velocity.

This larger flow will result in greater energy loss unless the system pressure is reduced. To help keep the actuator size to a minimum avoid excessive safety margins when calculating required actuator force.

Excessive actuator sizes will also increase the installed weight and the size of associated valves and pipework.

3.4 Valve Pressure Drop

Valve pressure drop is the difference between the supply pressure and the pressure drop across the actuator. Consideration of valve pressure drop may modify the selection of actuator size as described in 3.3 above.

The magnitude of valve pressure drop is a compromise between efficiency and control sensitivity. High efficiency requires low pressure drops whereas good sensitivity requires higher pressure drops across the valve.

A system that requires a high dynamic performance should use a higher valve pressure drop. This is because any change in valve pressure drop results in a change in valve flow gain and hence a corresponding change in system loop gain. Where the allowable loop gain is limited by stability considerations, it is useful to minimise the possible percentage change in valve pressure drop. A larger actuator area reduces the load pressure drop and this leaves a higher proportion of supply pressure available as valve pressure drop. This higher valve pressure drop will still change but by a smaller percentage.

A good compromise between efficiency and performance is obtained when the valve pressure drop is about two thirds of the supply pressure. There is little to be gained by increasing this further. Some advanced control techniques allow high dynamic performance at lower valve pressure drops. One such technique is to add an inner feedback loop using actuator velocity.

Open-loop systems often operate with lower valve pressure drops and use hydraulic methods to make the flow rate approximately independent of load pressure drop. This alternative method of obtaining consistent flow gain involves the use of a hydrostat to operate in conjunction with the electrohydraulic valve. Hydrostats are available which mount conveniently under the valve and provide meter in or meter out as required. These are not normally used for high performance closed-loop systems due to the speed of response of the hydrostat.

3.5 Valve Flow Rate

The product of the actuator area or motor displacement with the specified maximum velocity gives the required maximum flow rate. This should be matched as closely as possible to the valve maximum flow rate to minimise the effect of offsets and other valve non-linearities (see below). The quoted valve flow rate must be adjusted for the actual valve pressure drop occurring when maximum velocity is required.

3.6 Valve Dynamic Response

The dynamic response of the valve determines the transient response between input voltages (or current) and valve output flow. This, together with the dynamics of the load, determines the transient behaviour of an open-loop system. The load dynamics depend on several factors, the most important being load inertia, fluid compressibility, actuator area and oil volume.

In closed-loop systems stability requirements and the valve's dynamic response dictate the maximum loop gain. If the load inertia is low a faster valve will allow a higher system loop gain, that in turn helps to minimise the system's steady state error and improves the system's dynamic performance. If the inertia is high the load dynamics can dictate the maximum loop gain and a faster valve will not allow a higher loop gain. More information on calculating the valve's dynamic response for a particular system can be found in section 4 of this book.

3.7 Valve Null Characteristics

3.7.1 Null Position vs. De-energised Position

As stated in section 2.1 to 3 above the null position of the spool corresponds to zero input signal and zero flow. However, good dynamic performance normally requires a valve with a high pressure gain. The latter requirement may conflict with the need to know how the actuator will move if electrical power is lost.

The position of the spool when the valve is de-energised will depend on the construction of the valve. In some valves the spool is driven to the null position by preloaded springs. This means that some solenoid force is required to move the spool away from null but tolerances will usually allow the position defined by the spring to be slightly different to the true null.

In other valves the springs operating on the spool are arranged so that the net preload seen by the spool is zero. This helps the valve achieve good dynamic performance but allows the spool to move away from null with a small residual force in the solenoid. Alternatively, friction may stop the spool moving to the true null when the valve is de-energised.

A third style of valve uses a single spring to oppose the solenoid and this spring drives the spool to an offset position when the valve is de-energised. In this offset position the flow paths can be very different from those required at null and can be specifically arranged to obtain the required actuator movement. The disadvantage of the latter type of valve will be a short pulse of flow as the spool moves to the offset position.

If the system requires the actuator movement is minimised when the valve is de-energised it is possible to add pilot check valves between the valve and the actuator. These will be cross coupled so that they open and close automatically when required, as shown in figure 9.7.

3.7.2 Valve Lap Conditions

The lap condition of a valve affects the performance of the system particularly at small amplitudes.

A system that requires good accuracy and high dynamic performance will normally use a nominally zero lap valve. This is particularly important if the system is required to position the actuator to a high level of accuracy and may need to reverse if the actuator overshoots the required position. Zero lap conditions are usually combined with rectangular orifices to give substantially constant flow gain over the entire operating range. However, normal manufacturing tolerances affect the valve's flow gain in the null region which may typically result in a gain between 50% and 150% of nominal.

A valve with a low flow gain around null will need narrow metering notches in this region. This can reduce pressure gain unless the body and spool are made very accurately. High manufacturing tolerances can make a zero lapped spool underlapped. This results in a reduced pressure gain which in turn reduces the output stiffness.

Any overlap will introduce a delay when the actuator reverses. The magnitude of the delay will increase as the overlap increases. Overlap spools usually use preloaded springs so that the de-energised spool is positioned in the middle of the overlap zone to minimise leakage.

Overlap spools can have an intentional leakage to tank. The disadvantage of this is that the pressures each side of the actuator reduce when the spool is at null. This significantly reduces the stiffness of the actuator even if the system is not de-energised.

3.7.3 Pressure Gain

In a closed-loop system the actuator position error is a function of the valve pressure gain and the system's loop gain. The pressure gain determines the minimum input signal required to move the actuator and correct any position error. The electrical loop generates a signal to the valve based on the position error multiplied by the electrical gain.

3.7.4 Null Shifts

Valve null shifts are bias shifts which occur with temperature, supply and return line pressure changes and affect both open and closed-loop systems. They do result in a small but finite input signal to the valve being required to produce zero output from the valve. In a closed-loop system this signal is derived from the difference between the demand and output transducer signals and consequently results in a system error. The magnitude of the error decreases with increasing loop gain that in turn may be obtained by using a valve with higher dynamic response. To help minimise these errors the system can be designed to minimise pressure and temperature changes seen by the valve.

In applications where two or more systems are supplied by one power supply, large transients occurring in one system can cause rapid changes in both supply and return lines causing errors in the other systems. These can be reduced by using individual return lines and additional hysteresis in the supply.

Integral compensation can reduce the effect of null shifts. It can virtually eliminate the steady state error but its effect on the transient error may be more limited for rapidly occurring null shifts.

However, care must be taken to ensure that the use of integral compensations does not have an adverse effect on the stability of the loop. If too large a valve is selected for any application and only used over part of its stroke, the effect of null shifts hysteresis and overlap may be greater than expected.

3.7.5 De-energised Operation

Many valves will centre to the null position when de-energised and the speed of centring may produce unacceptable deceleration of the load and hence produce shocks in the system. Operation of the system if the valve is inadvertently de-energised should be considered.

Some valves offer a de-energised position which is not the null and alternative flow paths can be implemented to obtain specific hysteresis behaviour if the electrical drive to the valve is lost.

3.8 Linearity

In open-loop systems a low gain near to the null as shown in figure 2.12 is useful for providing repeatable low speed creep and helps to minimise the effects of tolerances on the valve overlap.

In closed-loop systems it is the loop gain that dictates system dynamics and this is a product of electrical gain and valve flow gain. When used in a system, a valve with a characteristic as shown in figure 2.12 will cause the system to have a variable gain that is lower when the valve is operating close to null. Consequently the system will have a lower dynamic performance when operating at small amplitudes. This undesirable characteristic can sometimes be reduced by introducing the inverse non-linearity into the electronic amplifier. However, if the valve transient response is slow the effect of the non-linearity is not so easy to eliminate. Also, if the non-linearities vary from valve to valve each amplifier may need adjustment to suit the valve it is driving.

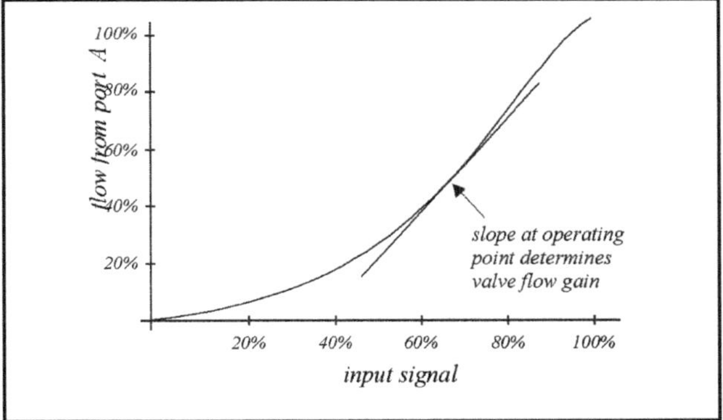

Figure 2.12: Effect of non-linear flow characteristic on gain

3.9 Hysteresis, Threshold and Resolution

Valve hysteresis, threshold and resolution are partly a result of internal friction within the valve. This causes the valve to move with a slip stick movement when subject to a slowly changing input signal. From a system perceptive, any hysteresis in the valve adds to the actuator position error.

An effective method of reducing hysteresis is to superimpose a dither signal onto the valve input signal. This signal can reduce the hysteresis both in the valve and in the actuator. However, the amplitude and frequency of the dither have to be carefully selected as excessive dither, particularly at low frequencies, can reduce the effective valve pressure gain and hence reduce the actuator position stiffness. Dither can also hide small overlaps and produce a slight increase in flow gain around null.

Another negative result of excessive dither is an audible and/or measurable movement of the actuator in response to the dither signal.

OPEN-LOOP SYSTEM CALCULATIONS

1. INTRODUCTION

In this chapter we present the equations for predicting the performance of a simple open-loop system consisting of a four-way electrohydraulic valve driving an actuator.

An open-loop system can be defined as a system in which the signal to the valve is not continuously modified in response to the measured system output.

Based on this definition we would normally expect a direct connection from the user input to the electrohydraulic valve. The user input can be a manual input or an electronic controller that outputs any sequence of predefined steps or ramps.

It is useful to note that the above definition includes the word "continuously". Consequently a system that interrupts the signal to the valve when the actuator reaches some predefined position can be designated an open-loop system. However, the signal applied to the valve after the interruption must consist of a pre-set ramp or step rather than any corrective action based on measurement of the final actuator position.

Typically, the sensors in such systems will only generate a switched output and this is used to initiate the ramp or step.

The direct connection between the user input and the valve dictates that the magnitude of the user input defines the velocity of the actuator.

1.1 Scope and Limitations

The calculations in this chapter ignore the dynamic aspects of the actuator motion. It is possible that the actuator may overshoot and/or oscillate when the user applies a step change to the input. Oscillation can often be avoided by applying a ramp to any signal to the valve. More information on allowable ramp times is presented in section 4.7 of this book.

NOMENCLATURE AND UNITS

Symbol	Description	Units
P_S	System pressure	N/m^2
ΔP_1	Valve pressure drop	N/m^2
ΔP_1	Valve pressure drop	N/m^2
A_1	Actuator area	m^2
A_2	Actuator area	m^2
F_X	External applied load	N

F_F	The actuator friction	N
F_T	Total actuator load	N
M	Load mass	kg
g	Acceleration due to gravity	m/sec ^{2}
S_A	Actuator extension or stroke	m
T_A	Time to complete actuator stroke	s
T_V	Response time of valve	(0-100%) s
Vmax	Maximum actuator velocity	m/s
ΔP_{valve}	Rated valve pressure drop	N/m ^{2}
Q	Rated valve flow	m 3 /s
θ	Angle of inclination	deg

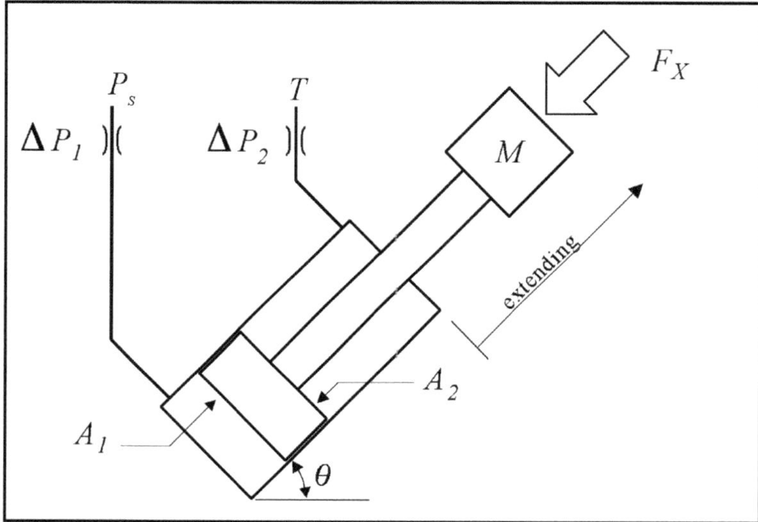

Figure 3.1: Actuator Configuration

2 CALCULATION METHOD

2.1 Calculate Net Force on Actuator

$$F_{T(\text{extending})} = F_X + F_F + Mg \times \sin(\theta) \tag{3.1}$$

$$F_{T(\text{retracting})} = F_X - F_F + Mg \times \sin(\theta) \tag{3.2}$$

Note that the actuator frictional force acts in the opposite direction to the direction of travel.

2.2 Calculate Required Supply Pressure

Check that the available supply pressure is sufficient to overcome the external load. If F_E is positive check:

$$P_S > \frac{F_{T(\text{extending})}}{A_1} \tag{3.3}$$

Also if F_R is negative check:

$$P_S > \frac{F_{T\,(\text{retracting})}}{A_2} \tag{3.4}$$

2.3 *Calculation of Valve Pressure Drops*

The following calculation assumes that the valve has equal flow metering areas in the two flow paths. The calculation simply finds the two valve pressure drops and these will remain constant regardless of actuator velocity provided that value of F is not velocity dependent.

2.3.1 Calculate valve pressure drop - Actuator extending

$$\Delta P_2 = \frac{P_S - \left[\dfrac{F_{T\,(\text{extending})}}{A_1}\right]}{\left[\dfrac{A_2}{A_1}\right] + \left[\dfrac{A_1}{A_2}\right]^2} \tag{3.5}$$

$$\Delta P_1 = \Delta P_2 \times \left[\frac{A_1}{A_2}\right]^2 \tag{3.6}$$

2.3.2 Calculate valve pressure drop - Actuator retracting

$$\Delta P_1 = \frac{P_S - \left[\dfrac{F_{T\,(\text{retracting})}}{A_2}\right]}{\left[\dfrac{A_1}{A_2}\right] + \left[\dfrac{A_2}{A_1}\right]^2} \tag{3.7}$$

$$\Delta P_2 = \Delta P_1 \times \left[\frac{A_2}{A_1}\right]^2 \tag{3.8}$$

2.4 *Check Valve Pressure Drop*

In order to provide a reasonable level of control over the load, at least 10% of the system pressure should be dropped across each effective orifice in the valve.

$$\Delta P_1 > 0.1 \times P_1 \tag{3.9}$$

$$\Delta P_2 > 0.1 \times P_1 \tag{3.10}$$

Check this condition for both extending and retracting cases.

If the load F_X changes, the actuator velocity will also change. The magnitude of the change can be minimised by increasing the percentage of system pressure dropped across the valve's two effective orifices.

2.5 Calculate Actuator Pressures

Actuator extending:

$$P_{1(\text{extending})} = P_S - \Delta P_{1(\text{extending})} \tag{3.11}$$

$$P_{2(\text{extending})} = \Delta P_{2\,(\text{extending})} \tag{3.12}$$

Actuator retracting:

$$P_{1(\text{retracting})} = \Delta P_{1(\text{retracting})} \tag{3.13}$$

$$P_{2(\text{retracting})} = P_s - \Delta P_{2(\text{retracting})} \tag{3.14}$$

2.6 Calculate Minimum Valve Rated Flow

To assist in the selection of a valve we will convert the valve flow and pressure drop as calculated above into the equivalent flow at the rated valve pressure drop of the valve under consideration. We will use the rated valve pressure drop for the total valve.

Firstly for the extending velocity.

$$Q_{(\text{extending})} = A_1 \times V_{\max} \sqrt{\frac{\text{Rated Valve Pressure Drop}}{2 \times \Delta P_{1(\text{extending})}}} \tag{3.15}$$

Secondly for the retracting velocity.

$$Q_{(\text{retracting})} = A_1 \times V_{\max} \sqrt{\frac{\text{Rated Valve Pressure Drop}}{2 \times \Delta P_{1(\text{retracting})}}} \tag{3.16}$$

2.7 Calculate Actuator Velocity for Maximum Valve Opening

Having selected a valve with more than sufficient flow it is possible to limit the velocity to the velocity defined in the system specification by limiting the input volts to the electrohydraulic valve.

$$\text{volts limit extending} = \frac{Q_{(\text{extending})}}{\text{Rated valve flow}} \tag{3.17}$$

$$\text{volts limit retracting} = \frac{Q_{(\text{retracting})}}{\text{Rated valve flow}} \tag{3.18}$$

3 WORKED EXAMPLE

Data

P_S	System pressure	210 bar = 210 x 10^5 N/m^2
D_1	Actuator bore	63 mm
D_2	Actuator rod	36 mm
F_X	External applied load	23 kN
F_F	The actuator friction	2 kN
M	Load mass	91 kg
V_{max}	Maximum Velocity	0.16 m/s
θ	Angle of actuator	0 deg
Q	Rated Flow of valve at 70 bar	20 L/min

Actuator area $A_1 = \dfrac{\pi \times 0.063^2}{4} = \mathbf{0.003117m^2}$

Actuator area $A_2 = \dfrac{\pi \times \left(0.063^2 - 0.036^2\right)}{4} = \mathbf{0.0021m^2}$

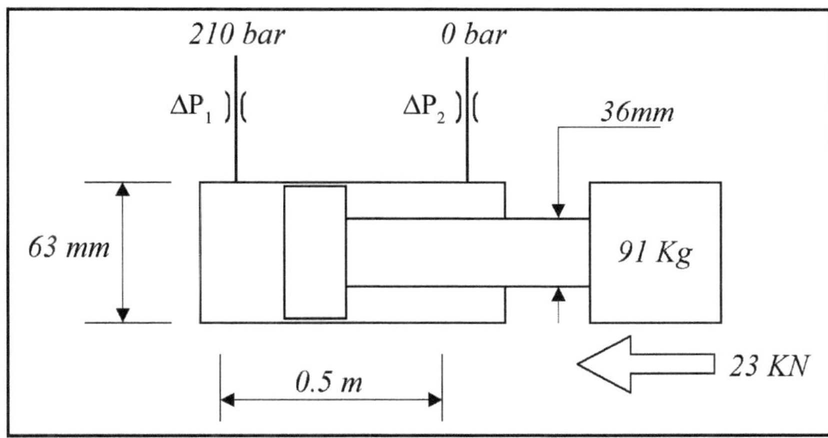

Figure 3.2: Example

3.1 Calculate Required Supply Pressure

$$P_S(\text{required}) = \frac{25\text{x}10^3}{0.003117} = \mathbf{80.2x10^5\ N/m^2} = 80.2\ \text{bar}$$

This is less that the 210 x 10^5 N/m^2 (210 bar) already selected.

F_T is positive and hence calculation of equation 3.4 is not required.

3.2 Calculation of Valve Pressure Drops

3.2.1 Calculate valve pressure drop - Actuator extending

$$\Delta P_2 = \frac{210 \times 10^5 - \dfrac{25 \times 10^3}{0.003117}}{\dfrac{0.0021}{0.003117} + \left[\dfrac{0.003117}{0.0021}\right]^2} = \mathbf{45.1 \times 10^5\ N/m^2} = 45.1\ \text{bar}$$

$$\Delta P_1 = 45.1\text{x}10^5 \times \left[\frac{0.003117}{0.0021}\right]^2 = \mathbf{99.4x10^5 N/m^2} = 99.4 \text{ bar}$$

3.2.2 Calculate valve pressure drop - Actuator retracting

$$\Delta P_1 = \frac{210 \times 10^5 + \dfrac{21 \times 10^3}{0.0021}}{\dfrac{0.003117}{0.0021} + \left[\dfrac{0.0021}{0.003117}\right]^2} = \mathbf{160 \times 10^5 N/m^2} = 160 \text{ bar}$$

$$\Delta P_2 = 160\text{x}10^5 \times \left[\frac{0.0021}{0.003117}\right]^2 = \mathbf{72.6x10^5 N/m^2} = 72.6 \text{ bar}$$

3.3 Check Valve Pressure Drop

Check that ΔP_1 and ΔP_2 are always at least 10% of the system pressure for both extending and retracting cases calculated above.

$$0.1P_S = 0.1 \times 210\text{x}10^5 = \mathbf{21 \times 10^5 N/m^2} = 21 \text{ bar}$$

3.4 Calculate Actuator Pressure

Actuator extending:

$$P_1 = 210\text{x}10^5 - 99.4\text{x}10^5 = \mathbf{110.6 \times 10^5 N/m^2} = 110.6 \text{ bar}$$

$$P_2 = \mathbf{45.1 \times 10^5 N/m^2} = 45.1 \text{ bar}$$

Actuator retracting:

$$P_1 = \mathbf{160 \times 10^5 N/m^2} = 160 \text{ bar}$$

$$P_2 = 210\text{x}10^5 - 72.6 \times 10^5 = \mathbf{137.4 \times 10^5 N/m^2} = 134.7 \text{ bar}$$

3.5 Calculate Minimum Valve Rated Flow

Assuming the maximum velocity is for the extending actuator.

$$Q = 0.003117 \times 0.16 \times \sqrt{\frac{70 \times 10^5}{2 \times 99.4 \times 10^5}} = 0.0003\text{m}^3/\text{sec} = \mathbf{17.7 \text{ L/min}}$$

And for the retracting actuator

$$Q = 0.003117 \times 0.16 \times \sqrt{\frac{70 \times 10^5}{2 \times 160 \times 10^5}} = 0.00023\text{m}^3/\text{sec} = \mathbf{14 \text{ L/min}}$$

3.6 Calculate Maximum Valve Input Signal

To calculate the voltage required to reach the maximum velocity we will assume that the valve input voltage required to reach the valve's rated flow in each direction is ±10 volts.

$$\text{volts limit extending} = \frac{17.7 \times 10}{20} = 8.85 \text{ V}$$

$$\text{volts limit retracting} = \frac{14 \times -10}{20} = -7.5 \text{ V}$$

PERFORMANCE PREDICTION OF CLOSED-LOOP SYSTEMS

1 INTRODUCTION

In this chapter we describe a simple method of predicting the performance of a closed-loop electrohydraulic position control system as shown in figure 4.1. Such systems allows the user to continuously vary the position of an actuator using an electrical input signal to define the desired position. In operation the system automatically generates the correct drive signals for the electrohydraulic valve so that the actuator is driven to the desired position.

A particularly useful feature of the prediction method described in this chapter is that the design process starts with a fully operational system, hereafter called the framework system. This framework system is operational because it is populated with imaginary components whose internal functions and interface signals meet the requirements of the system.

The task of the system designer is simply to input the real system requirements and then to replace the imaginary components with real components. This replacement or substitution process can be performed without loss of overall performance provided all the new components have the same function and interface as the component they replace. In this way the new system will always be operational and the final design should conform to the original system requirements.

Rather than just presenting the equations we explain the operation of the system and how the equations are derived. An additional simplification can be used if the mass attached to the actuator is light. In these circumstances, some basic performance predictions can be made using just mental arithmetic. However, a general method that is equally applicable to light and heavy loads is also presented.

It will be shown later in this chapter that the method can be used for both digital and analogue systems.

It may be of interest to note that a closed-loop system can control actuator position, velocity or even force, provided the system has the sensors to measure the desired parameter. However, the discussion in this chapter is limited to closed-loop position systems.

1.1 Scope and Limitations

The method is applicable to systems consisting of the elements shown in figure 4.2. All the components are assumed to have a linear relationship between their input and output. This includes the hydraulic flow gain of the valve and the electrical gain. However, the method is a good starting point for predicting the performance of systems with small non-linearities.

The method predicts the worst case position error rather than the system repeatability. The latter may be significantly better than the worst case position error but the magnitude is more difficult to calculate and less dependable.

The extra simple method is only applicable to lightly loaded systems.

1.2 Overview of the Design Process

The method starts with a fully operational system populated with imaginary components shown as blocks in figure 4.2. During the design process it will be necessary to replace the imaginary component with real components to create the operational system. This can be done, undone and redone as often as required provided the interface and function of the block remains unchanged. If the interface is not the same, an imaginary interface card must be added to convert the interface of the real component so that it matches the interface of the block being replaced. The interface card will change the relationship between the component's parameters and the overall system performance. Consequently the substitution rules in section 6 of this chapter must be followed when imaginary interface cards are used during the design.

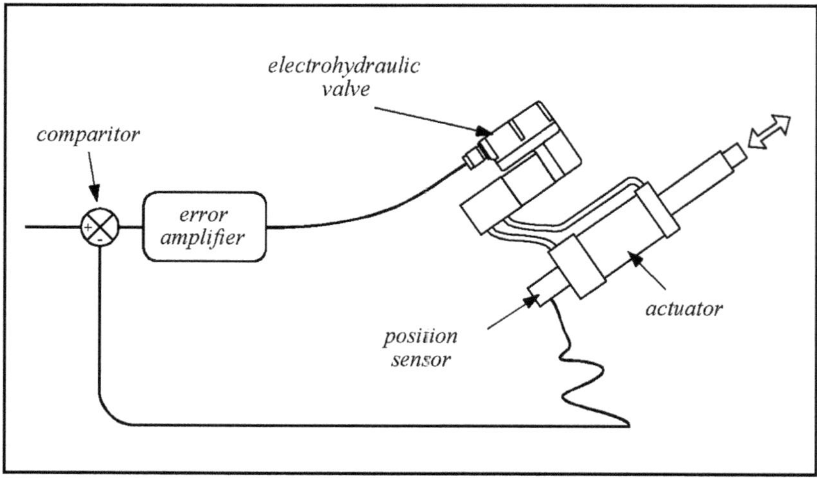

Figure 4.1: Typical System

NOMENCLATURE AND UNITS

Additions to nomenclature given in part 3.

Symbol	**Description**	**Units**
ΔP_E	Pressure difference between valve ports (extending)	N/m^2
ΔP_R	Pressure difference between valve ports (retracting)	N/m^2
V_1	External pipe volume attached to actuator area A_1	m^3
V_2	External pipe volume attached to actuator area A_2	m^3
ω_m	Oil spring natural frequency	Hz
ω_{valve}	Valve 90_ phase lag frequency	Hz

S	System stiffness	N/mm
K_E	Electrical gain	none
K	Estimated loop gain	s^{-1}
X_u	Estimated maximum position error	mm
X_f	Estimated maximum following error	mm
V_X	Maximum actuator velocity (extending)	m/s
V_R	Maximum actuator velocity (retracting)	m/s
ΔP_{valve}	Rated valve pressure drop	N/m^2
Q	Rated valve flow	m^3/s
S_c	Estimated closed-loop stiffness	N/mm

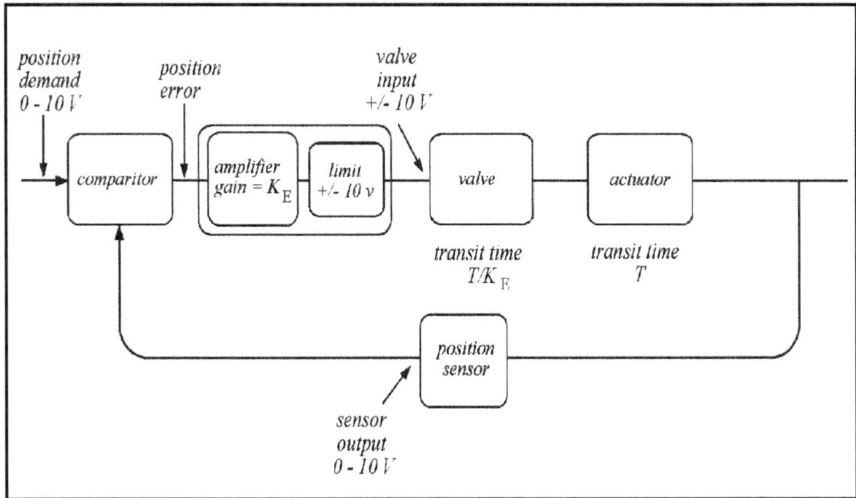

Figure 4.2: System Components

2 SYSTEM COMPONENTS AND THEIR OPERATION

The framework system shown in figure 4.2 contains six components, shown as blocks. The component parameters and interfaces shown have been selected to make the equations as simple as possible. In some cases the operating parameters of the components do not have absolute values but are specified relative to other system parameters.

The important characteristics of these blocks are now described.

2.1 The Electrohydraulic Valve

The electrohydraulic valve block represents the electrohydraulic valve and its amplifier. We do not need to know the details of any signals between the two or whether the amplifier is mounted on the valve or mounted remotely. For simplicity we will, from here on, refer to the input to this block as the input to the valve.

The framework system does not define the type of valve to be used. However, from experience, we know that the traditional four-way, zero lapped, spool valve is ideally suited to controlling an actuator.

The input signal to the valve is always in the range of ±10V. When the input signal is zero the output flow from the valve is assumed to be zero. In this ideal valve the flow metering areas increase linearly to a maximum at ±10V. The valve is able to drive the actuator in both directions such that positive volts produces positive actuator velocity and negative volts produces negative actuator velocity.

The four valve performance parameters considered when calculating system performance are the step response time, maximum flow, hysteresis and pressure gain.

The valve step response time is designated T_V and is the time to move from 0 to 10 volts, i.e. null to fully open, or fully open to closed, whichever is the longer. The real valve used in this system must not overshoot at the end of a step. If the valve exhibits overshoot it may be possible to apply a ramp or slew rate limit within the valve block to eliminate the valve overshoot. If this solution is used the ramp must remain in the final system and T_V will include the ramp time.

The maximum output flow from the valve is not specified in the initial framework system but is assumed to be exactly equal to that required to drive the actuator at the calculated maximum velocity in each direction. The design method will make corrections for any mismatch between the desired flow and the actual flow of the selected valve.

Valve hysteresis h_V is specified in volts at the input to the valve block and equals the deviation from the ideal steady state relationship between input volts and spool position. The deviation produced by the hysteresis does disappear with time and has the opposite sign when the same position is approached from the opposite end of spool travel. Hysteresis is specified as $\pm h_V$ in volts.

In a perfect zero lapped valve the pressures at the two service ports would be P_S and P_T and these pressures would reverse instantaneously whenever the input signal passed through zero. The imperfections in the valve give rise to a slope in this pressure reversal and this slope is known as the pressure gain. The pressure gain G_V is measures in bar/volt where the pressure is the difference between the two service port pressures. Alternatively, the pressure gain can also be recorded as a plot showing the two individual service port pressures against valve input signal.

2.2 The Actuator

The actuator used in the framework system can be a linear cylinder or a rotary actuator. If a rotary actuator is used any references to position should be replaced by a reference to angular position. These actuators can be symmetrical or asymmetrical. An asymmetrical actuator is any actuator in which the flow in is not equal to the flow out.

The actuator parameters considered in the framework system are the desired actuator transit time, the actuator stroke, the area on each side of the piston and the combination of the static loads on the actuator and any friction forces. The predicted actuator position error is usually an output from the calculation but if the maximum allowable error is part of the specification this error can be used to calculate other system parameters.

For the framework system the transit time is designated T_A while the stroke is designated S_A.

The size of the actuator may be selected to suit the loads applied to the actuator. Alternatively, we may use an existing actuator and test its potential performance in the system. In either case the combination of the actuator size and the valve's maximum flow are assumed to give the required actuator transit time.

When an asymmetrical actuator is selected it is usual for the calculations to predict a different velocity for the two directions of movement. The framework system deals with this situation by forcing the system to operate at the lower velocity in both directions. This topic will be addressed in more detail later.

Usually the real actuator will have a stroke slightly longer than S_A to avoid impact with the end stops.

2.3 The Position Sensor

The framework system does not define the type of position sensor but simply specifies that the sensor provides an electrical signal proportional to the actuator's linear or angular position.

The sensor always gives an output of 0 to 10V over the stroke S_A but in the real system we will want to avoid damage to the sensor by making sure that the maximum mechanical stroke of the sensor is slightly greater than the mechanical stroke of the actuator. Also to maintain correct system operation during overshoots the sensor must provide a linear output over the full range of actuator travel. Consequently the imaginary sensor must have an output range in excess of 0 to 10v.

The sensor parameters which can influence system performance are resolution, linearity and response time. Normally the sensor will have a response time ten times faster than the system response time such that this latter parameter can be ignored.

The sensor is arranged so that the actuator moves towards the 10V position when the input to the electrohydraulic valve is positive.

2.4 The Position Comparator

This block generates an output signal that is equal to the difference between the desired actuator position and the actual actuator position.

The desired actuator position is a user supplied input signal that defines the desired position of the actuator. In the framework system this signal is limited to a range of 0 to 10 volts.

The output is the position error signal and this is generated by subtracting the actual position signal from the desired position signal. Consequently the error signal has range of ±10V.

The substitution rules allow the function inside this block to be performed by analogue or digital means. Typically there will be no errors associated with an analogue version of this block but if implemented as a digital process then we should assume the system can have an additional error equal to one least significant bit.

2.5 The Electrical Gain and Limiter Block

The input signal to this block is output from the comparator and consequently is within the range of ±10V. The gain of this block is the ratio between the output and the input. For example, when the gain is equal to 4 the maximum output from the gain block will be ±40V. Clearly this voltage is too great for the valve and a limiting function is always required. The default output limits will be set to ±10v.

The gain of the block is designated K_E and this parameter is the focus of most of the design effort because it determines both the dynamic response and the final position accuracy. For systems with low actuator loads the gain K_E can be calculated from equation 4.1 where T_A and T_V are the response times of the actuator and valve as described above.

$$K_{E1} = \frac{T_A}{T_V} \tag{4.1}$$

Note that we have added the suffix 1 to K_E to designate which equation was used to calculate the gain.

The gain block will have two additional parameters. The positive clamp and the negative clamp voltage.

Note that K_E, the electrical gain of this block, is not the same as the closed-loop gain as the latter is the product of all the gains around the closed-loop of the system. We will only use K_E in this method.

3 THE TRAPEZOIDAL VELOCITY PROFILE

The simplified method of predicting performance described in this chapter is possible because the movement of a lightly loaded actuator will approximate to the trapezoidal velocity profile shown in figure 4.3. This is not a "user supplied" profile but is automatically generated by the operation of the closed-loop system.

The movement can be divided into three phases: acceleration, constant velocity and deceleration. The acceleration depends on the rate of opening of the valve. The constant velocity depends on the maximum opening of the valve. The deceleration depends on the rate at which the valve closes. We will see later that the third phase also depends on electrical gain and the speed of response of the valve.

To describe the trapezoidal profile we will assume the system is subject to a step input from 0 to 10 volts.

3.1 The First Phase

The first phase starts with the actuator stationary at one end of the stroke. We will assume the actuator is at the 0V position and the demand input is at 0V. As the actuator position error is zero the input signal to the valve is also 0V. The user input signal then steps to +10V. This produces a 10V step at the output of the comparator. This 10 volt signal is amplified by the electrical gain producing a output signal K_E times bigger. However, additional circuits limit this signal so that during the first phase the input to the valve is a constant +10 volts.

The electrohydraulic valve cannot respond instantaneously to this step input but will open as fast as it can. The perfect imaginary valve is assumed to open and close in the same time so that the trapezoidal waveform is symmetrical. If the oil compressibility can be ignored the actuator velocity will be proportional to the valve flow and hence the actuator will ramp from zero to maximum velocity in direct response to the valve opening. This forms the first part of the trapezoidal profile.

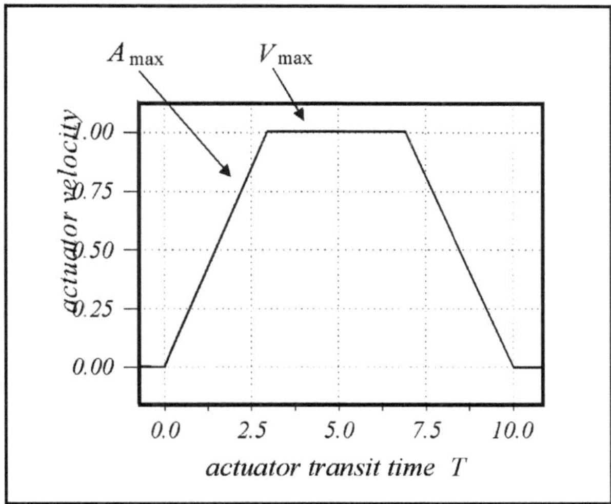

Figure 4.3: The Trapezoidal Waveform

During the first phase the actuator is assumed to have constant acceleration A_{max} that can be calculated from the trapezoidal waveform using the stoke S_A, the required transit time T_A and the electrical gain K_E equation 4.2.

$$A_{max} = \frac{K_E \times (1 + 2/K_E)^2 \times S_A}{2 \times T_A^2} \tag{4.2}$$

3.2 The Second Phase

During the second phase the actuator moves towards the demand position at the maximum velocity V_{max}. The maximum actuator velocity required during the second phase can be calculated from the trapezoidal waveform as shown in equation 4.3 .

$$V_{max} = \frac{(1 + 2/K_E) \times S_A}{T_A} \tag{4.3}$$

The position error is continuously reducing during this phase but electrical error signal multiplied by the electrical gain remains in excess of 10 volts. As a result, the input to the electrohydraulic valve is held at +10 volts and the valve is held fully open. Consequently the actuator velocity V_{max} is dictated by the supply pressure, the valve pressure drops and the actuator loads. This equates to operating in an open-loop mode and the equations presented in chapter 3 can be used to calculate the required valve flow rating.

3.3 The Third Phase

The third and final phase of the trapezoidal profile is the most important phase as it defines the maximum electrical gain. The start of this phase is defined as the moment when the input to the valves falls below +10 volts and hence the electrical circuits for limiting the signal are not operative.

When the system enters the third phase the actuator is moving at its maximum velocity and as a result the error signal is also reducing at a maximum rate. The framework system has an electrical gain of α and consequently the input to the valve will start to fall when the remaining actuator stroke is S_A / K_E. For example, if the electrical gain K_E is 4 the remaining stroke is 25% of S_A.

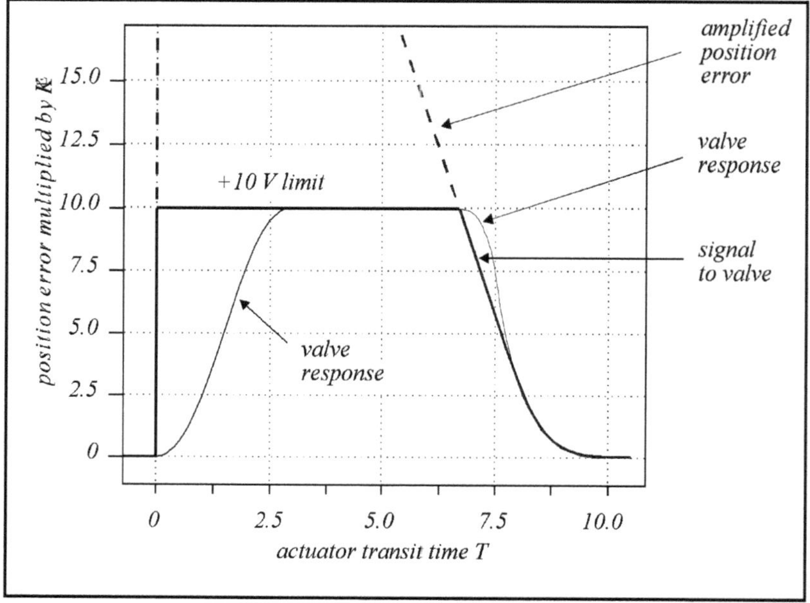

Figure 4.4: Trapezoidal Response

The valve has no advance warning that the input to the valve is about to fall below 10V and hence the valve will take a little time to respond to this sudden change. To avoid an overshoot in the final actuator position the valve must be able to catch up with this rapidly changing input signal and then follow the input signal so that the valve closes at the exact moment the actuator reaches its final position.

Throughout this phase the valve is closing and hence the actuator velocity will reduce. This reduction in actuator velocity extends the time available for the valve to catch-up as shown in figure 4.4. However, the valve still has to be able to perform the initial catch-up and consequently we specify that the valve closing time T_V must equal T_A / K_E. This is the main framework system equation.

At the end of this phase the ideal valve will reach its zero flow position at exactly the same time as the actuator error reaches zero.

4 LIMITS ON MAXIMUM GAIN DUE TO COMPRESSIBILITY

The method described above requires that the system follows a trapezoidal waveform. This waveform can only be generated if the actuator velocity tracks the valve flow. Failure to maintain this relationship will result in system overshoot because the actuator will not stop at the exact moment the valve closes.

The main cause of this problem is oil compressibility. Compressibility becomes more significant if the oil volumes are high or the actuator loads are high. Fortunately the effect of compressibility can be reduced if actuator acceleration forces are reduced.

Making acceleration less means the third phase must be proportionally longer and/or the actuator transit time must be increased. This clearly impacts on the electrical gain K_E and an alternative method of calculation is required.

The method proposed in this chapter is based on the natural frequency of the actuator and the actuator transit time. If the new gain is lower than the gain calculated using equation 1 we will use this new lower gain.

4.1 Calculate Oil Spring Stiffness

The two oil volumes on each side of the piston act as independent springs and the stiffness due to each must be summed to find the net actuator stiffness. Technically we need to find the stroke position with the lowest stiffness and use this value to calculate the natural frequency of the actuator. However, in the following equation we will find the stiffness at mid stroke.

$$St = \frac{A_1^2 * 10^9}{V_1 + [A_1 * 0.5 * S_A]} + \frac{A_2^2 * 10^9}{V_2 + [A_2 * 0.5 * S_A]} \tag{4.4}$$

4.2 Calculate Oil Spring Natural Frequency

$$\omega_n = \sqrt{\frac{S_t}{M}} \tag{4.5}$$

$$K_{E2} = \frac{\omega_m * T_A}{10000} - 2 \tag{4.6}$$

4.3 Calculate a New Electrical Gain Based on Actuator Stiffness

$$K_{E2} = \frac{\omega_m * T_A \sec}{10} - 2 \tag{4.7}$$

As T is in milliseconds we can use the easier to remember alternative equation 4.7.

Note that we have designated this gain as K_{E2} to indicate the alternative method of calculation.

4.4 Select the Lower Gain

Compare K_{E1} with K_{E2} and select the lower.

5 ACTUATOR POSITION ERRORS

In an ideal system the actuator will continue to move at ever decreasing speed until the actuator error reaches zero at exactly the same time as the input to the valve reaches zero volts.

Unfortunately the construction details in any electrohydraulic valve will prevent it having an infinite pressure gain and consequently the pressure available to move the actuator will reduce as the valve approaches null in accordance with the pressure gain characteristic. This reduction of pressure will allow the actuator to stop moving before it reaches the desired position. However, any error that exists at the end of a step will be eliminated by the system if the system can start the actuator moving with the same or less error. This logic means that we only need to investigate the maximum steady state error that can exist while the actuator is stationary.

To do this we assume the actuator is stationary and the user increases the error by making a small change to the demand input signal. Increasing the error signal automatically increases the input signal to the valve. This in turn increases the pressure difference seen at the valve service ports but the actuator will not move unless the pressure difference generated by the valve is sufficient to overcome the forces acting on the actuator.

We will also assume the valve used in the system has a balanced lap condition. This implies the mean pressure at the service ports is 50% of the difference between supply pressure and tank pressure. This is used to simplify the calculation of the differential pressures ΔP_E and ΔP_R for extension and retraction respectively.

5.1 Calculate Pressure to Start Actuator Motion

Equation 8 uses the conventions shown in figure 3.1 to define the sign of the force F and assumes a positive ΔP is generated when P_1 is greater than P_2. This equation is shown for the extending case but the same equation is used for motion in both directions. Remember the static friction should be included in F and the static friction will always acting to oppose the intended direction of movement.

$$\Delta P_E = \frac{2F_E - (P_S - P_T)(A_1 - A_2)}{(A_1 + A_2)} \tag{4.8}$$

Note that the differential pressures ΔP_E and ΔP_R should be less than 80% of the supply pressure. Beyond this range the pressure gain is likely to be substantially reduced. If this is not achieved consider a higher supply pressure or larger actuator piston areas.

5.2 Calculate Valve Input Volts

We must now calculate the volts at the input to the valve necessary to develop the pressure difference calculated in equation 7. This uses a single numerical value for valve pressure gain.

$$V_E = (\Delta P_E \,/\, PressureGain_V) + h_V \tag{4.9}$$

$$V_R = (\Delta P_R \,/\, PressureGain_V) - h_V \tag{4.10}$$

Remember to carry the sign of ΔP through to equations 4.9 and 4.10. Note that static friction in the actuator and valve hysteresis h_V will both increase the difference between V_E and V_R.

To find the corresponding actuator position error in volts we simply divide V_E and V_R by the electrical gain K_E. This error can be converted into a physical distance by multiplying by the stroke S_A and dividing by 10 as shown in equation 4.11. The units of X_E will be the same as S_A and V_E will always be expressed in volts. Equation 4.11 is used for both extending and retracting cases.

$$X_E = \frac{V_E}{K_E} * \frac{S_A}{10} \tag{4.11}$$

We now have two errors, one for each direction of travel. These errors may be of the same sign or of opposite sign. In either case the errors may have a non-zero mean. It is common for any non-zero mean to be eliminated during commissioning using a bias signal or by adjusting the physical position of the actuator position sensor.

After such commissioning the system will appear to have a zero mean error and the scatter about this mean will have a magnitude equal to half of the difference between X_E and X_R.

6 FOLLOWING ERROR

The following error is a measure of the system's ability to follow a voltage ramp applied to the input. If the actuator follows the ramp it will move with constant velocity. The following error is defined as the steady state difference between the desired position and the actual actuator position while the system is moving. The following error can only be calculated if the velocity required to follow the ramp is less than the maximum system velocity.

The calculation of following error is very simple. If, for example, we input a ramp that equates to a velocity equal to half the maximum velocity, the input to the valve will have to be 5 volts. To generate 5 volts at the input to the valve the output from the comparator must be 5 divided by K_E. This can be converted into a physical distance by multiplying by the stroke S_A and dividing by 10. This can be rearranged into equation 4.12 where *Velocity Ratio* is the actual speed divided by V_{max}.

$$Following\ Error = \frac{Velocity\ Ratio \times S_A}{K_E} \tag{4.12}$$

It will be seen that the following error reduces as the speed reduces. At low speed we should make an allowance for valve pressure gain and hysteresis by adding the maximum static actuator position error to any following error estimate.

7 THE SUBSTITUTION RULES

During system design the contents of any block in the framework system can be changed provided the substitution rules detailed below are followed. By following these rules the system designer can easily incorporate components with non-standard interfaces and test them without re-designing the total system.

The use of standard interfaces allows the substitution process to be undone and redone as many times as desired.

The substitution rules are listed below:

1. The internal operation and the external interfaces of any complete block must remain the same as the original definition for that block.
2. The component within any block can be changed provided it can be made to mimic the component it replaces. To convert any electrical interface signals to the correct magnitude and sign the new component may be accompanied by an imaginary electrical adapter card.
3. The imaginary valve can be replaced by a real valve provided the flow is greater than or equal to the required flow. To obtain the required flow in both either directions an attenuator is added in front of the valve but within the valve block. An attenuator is used rather than a clamp as the flow output from the valve must vary linearly from zero to the required flow over the complete input range of 0 to ±10 volts.

   In equation 4.13 the rated pressure drop for a single flow path is divided by calculated valve pressure drop for the same flow path. Consequently the factor 0.5 is included because we have assumed the valve rated pressure drop is from P to T.

$$V_{reduced} = 10 * \frac{requiredflow_{PA}}{valveflow(rated)}\sqrt{\frac{0.5 * valvepressuredrop(rated)}{calculatedvalvepressuredrop_{PA}}} \qquad (4.13)$$

   The above equation is used for the extending case and for the retracting case the suffix *PA* is replaced by *AT*. Usually the two calculated voltages will be different and consequently any attenuator is likely to require different gains for positive and negative voltages. This is the preferred method of equalising the velocity in each direction. To be sure of good performance the designer may wish to check that change in gain takes place while the actuator is stationary.
4. If an attenuator is used within the valve block to obtain the desired maximum velocity the apparent hysteresis of the valve will increases by the inverse of the attenuator gain. This may result in different values of positive and negative h_V if the attenuator gains are different. Also the value of the pressure gain used in any calculations is reduced by the value of the attenuator.
5. When the design is complete the attenuator can be combined with the electrical gain to create a single gain block. If this is done the clamps at the input to the real valve will be set at the reduced voltage rather than the default ±10 volts. Similarly, other elements can be moved around using the normal rules applying to block diagrams. However, the elements in the framework system used for predicting system performance should not be merged as the equations depend on the fixed interfaces shown in figure 4.2.

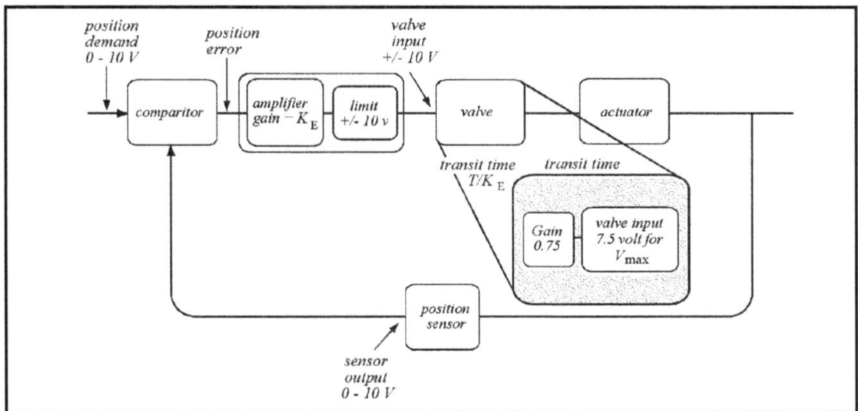

Figure 4.5: Substitution of Valve with Valve and Attenuator

8 OBSERVATIONS REGARDING OPEN-LOOP SYSTEMS

It is of interest to note that the calculation of maximum loop gain α as described above is the only system parameter we have to consider when calculating where in the stroke of the actuator the valve starts to close. For example, a system with an electrical gain of 5 will always start closing the valve when the actuator reaches the last 1/5 of its stroke.

We can also assume that the lower value of gain α_2 calculated when allowing for actuator mass and fluid compressibility defines a rate of valve closing that will not provoke overshoots or actuator oscillation.

Combining these observations we can make the approximate calculation that to avoid overshoot the time to close the valve in an open-loop system should be limited to T_A/α_2. This can be done by the addition of a ramp circuit in the drive to the valve.

9 WORKED EXAMPLE

In this example we assume the system designer knows the desired actuator response time, the applied loads, the actuator size and the valve characteristics. This will allow us to calculate the electrical gain and the expected position accuracy.

For simplicity we will use an example system similar to that presented in part 3.

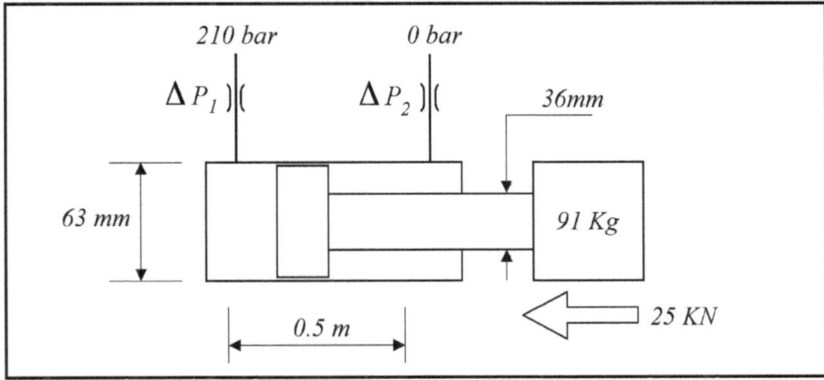

Figure 4.6: Example

Data

System pressure		210 bar = 210 x 10^5 N/m^2
Tank pressure		0.0 bar
Actuator stroke		0.5 m
Actuator bore		63 mm
Actuator rod		36 mm
Actuator Orientation		Horizontal
Actuator Friction		4700 N
External applied load		25 kN
Load mass		91 kg
Actuator Response Time		3.125 sec
External pipe volume	V_1	1 litre = 1 x 10^{-3} m^3
External pipe volume	V_2	1 litre = 1 x 10^{-3} m^3
Velocity		0.16 m/s
Valve Step Response		7.5 ms
Valve Pressure Gain		400 bar/volt**
Valve Hysteresis		±0.06volts
Valve Maximum Flow		25 L/min at 70 bar

** measured with 210 bar supply pressure

The following calculations follow the equations 1 to 11 detailed above.

9.1 Maximum Electrical Gain

From equation 4.1

$$K_{E1} = \frac{T_A}{T_V} = \frac{3125}{7.5} = 416 \quad volts/volt$$

9.2 Calculate Oil Stiffness

From equation 4.4

$$St = \frac{0.003117^2 * 10^9}{0.001 + [0.003117 * 0.5 * 0.5]} + \frac{0.0021^2 * 10^9}{0.001 + [0.0021 * 0.5 * 0.5]} = 8.352 * 10^6 \text{ N/m}^2$$

9.3 Calculate Actuator Natural Frequency

From equation 4.5

$$\omega_n = \sqrt{\frac{8.352 * 10^6}{91}} = 303 \, rad/\sec$$

9.4 Calculate a New Electrical Gain

From equation 4.7.

$$K_{E2} = \frac{303 * 3.125}{10} - 2 = 93 \quad volts/volt$$

In this case we will use the lower gain of 93.

9.5 Calculate Maximum Velocity

For completeness we will include the calculation of maximum velocity as shown in equation 4.3. However, for high gains the maximum velocity is almost identical to the average velocity.

$$V_{\max} = \frac{(1 + 2/93)*0.5}{3.125} = 0.163 m/\sec$$

9.6 Maximum Flow

Next we calculate the maximum flow by multiplying the velocity by actuator area A_1.

$$Flow_{MAX} = 0.163 * 0.003117 = 0.0051 m^3/\sec = 30.4\, l/\min$$

9.7 Calculate Volts to Produce Maximum Flow

The valve must be capable of this flow at the pressure drop of 87.9 bar as calculated in chapter 5. Equation 4.13 is used to find the valve input volts for both the extending and the retracting case.

$$V_{EXTENDING} = 10 * \frac{30.4}{25}\sqrt{\frac{0.5*70}{87.9}} = 7.67\, volts$$

For the retracting case the flow stays the same because the velocity is the same. The difference is that the actuator pressures change as the pressure intensification that existed during extension does not exist during retraction.

$$V_{RETERACTING} = 10 * \frac{30.4}{25}\sqrt{\frac{0.5*70}{158.2}} = 5.72\, volts$$

9.8 Calculate the Attenuator Gains and Modified Valve Parameters

Positive Attenuator = 7.67 volts/10 volts = 0.767

Negative Attenuator = 5.72 volts/10 volts = 0.572

Positive Hysteresis = h_V/ 0.767 = 0.06/0.767 = 0.08 volts

Negative Hysteresis = h_V/ 0.572 = 0.06/0.572 = 0.10 volts

Positive Pressure Gain = $Pressure\ Gain_V$* 0.767 = 400 * 0.767 = 307 bar/volt

Negative Pressure Gain = $Pressure\ Gai_{nV}$ * 0.572 = 400 * 0.572 = 229 bar/volt

9.9 Calculate the Required Pressure Difference

Calculate the system pressure required to support the applied static loads using equation 4.6 and assuming a balanced valve lap condition.

$$\Delta P_E = \frac{2(25000 + 4700) - (210*10^5)(0.003117 - 0.0021)}{(0.003117 + 0.0021)} = 72.92 \text{ bar}$$

And the same for the retracting actuator.

$$\Delta P_R = \frac{2(25000-4700)-(210*10^5)(0.003117-0.0021)}{(0.003117+0.0021)} = 36.88 \text{ bar}$$

9.10 Calculate Voltage Required to Start Actuator Motion

Using Equation 4.9 and 4.10

$$V_E = 72.92/307 + 0.08 = 0.317 \; volts$$

$$V_R = 36.88/229 - 0.10 = 0.061 \; volts$$

9.11 Calculate the Maximum Actuator Errors for each Direction

Note that the original calculated value of α is used equation 4.11 below as h_V and pressure gain have already been modified before being used in equations 4.9 and 4.10.

$$X_E = \frac{0.317}{93} * \frac{0.5}{10} = 0.170\text{mm}$$

$$X_R = \frac{0.061}{93} * \frac{0.5}{10} = 0.033\text{mm}$$

9.12 Calculate the Error Amplitude after Commissioning

$$Error\ Amplitude = 0.5*(0.173 - 0.033) = \pm 0.1 \; mm$$

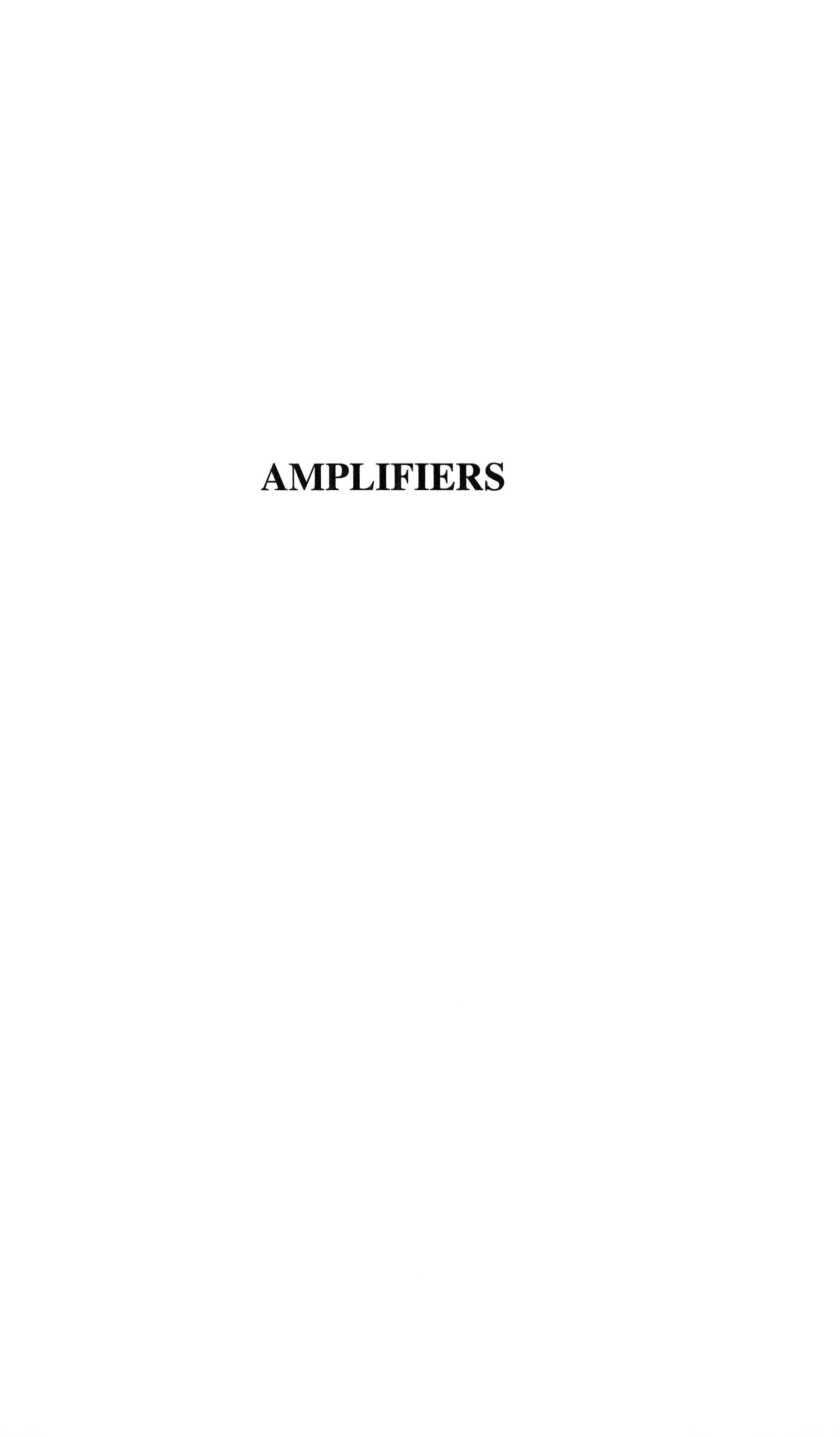

AMPLIFIERS

1 DRIVE AMPLIFIERS

The interface between the system electronics and the electrohydraulic valve is usually an electronic amplifier. A good amplifier can be of considerable assistance in building a complete system if it provides additional functions such as ramps to reduce shocks by smoothing out steps in the input signal or provides circuits to move the spool quickly through the valve's overlap zone.

To obtain maximum performance the drive amplifier should be compatible with the electrohydraulic valve and this selection can be simplified by using an amplifier recommended by the valve's manufacturer. The characteristics of the input and output circuits should match the requirements of the valves, transducers and actuators to avoid damage to the valve or the amplifier. If the valve has a position or pressure transducer the matching of amplifier to valve is particularly important.

The following items describe various aspects of amplifier performance which should be considered when selecting an amplifier for electrohydraulic valves.

2 AMPLIFIER FORMATS

2.1 Eurocard Amplifiers

The majority of amplifiers for electrohydraulic proportional valves are available on Eurocard sized (100 x 160 mm) printed circuit boards and use a two-part edge connector. Suitable card holders are available from a large number of suppliers and rack mounting card holders are particularly useful when multiple Eurocards are being used. The additional Eurocards may be system control cards or other valve amplifier cards. A metal card housing can provide shielding to RFI and EMI when required.

2.2 Valve-mounted Amplifiers

The principal advantage of valve-mounted amplifiers is the reduction in wiring needed to install the valve, resulting in quicker and easier installation with less opportunity of wiring errors and less chance of accidental loss of feedback signals during operation. In addition, the amplifier can be factory matched to the valve providing consistent performance valve to valve.

The disadvantage of valve-mounted amplifiers is that the electronics are subject to the same environmental conditions as the valve. High quality electronics can survive high levels of shock and wide temperature ranges but each potential application needs to be checked carefully.

Some special versions of valve-mounted amplifiers can ease the replacement of a switching valve with an electrohydraulic valve. These special amplifiers allow the use of existing wiring and system control equipment to provide power and timing signals to the valve while the amplifier adds the special features such as ramps, required by the electrohydraulic valve.

3 ENVIRONMENT

3.1 Moisture, Humidity and Dust

Clearly some applications will require more environmental protection than others. Good environmental protection will increase the cost of the amplifier and may require more careful and costly installation procedures such as the use of multicore cable and cable glands for all signals entering the amplifier. Where connectors are exposed to the environment both mating parts must be to the required standard.

The most commonly specified standard for protection against water splashes is IP 64 as specified in IEC 60529 (BS EN 60529). This standard provides for an environment totally protected against dust and water splashes. It also provides moderate protection against low-pressure water jets with some limited ingress being permitted.

In very damp, but not immersed, applications protection to IP 66 should be specified. For immersed conditions or where high pressure water jets will be used IP 67 will be necessary. If the amplifier is to be mounted inside a larger box or wiring cabinet, environmental protection is seldom required provided the housing has an adequate IP rating.

A unit that is subject to changing temperatures will experience changes of internal pressure as the air inside the unit expands and contracts. If, at the same time, the unit is subject to high humidity the unit may become damp inside if water vapour is drawn in past any seals or joints in the housing. Environmental protection for such conditions is covered in IEC 68.

3.2 Vibration

The vibration levels applied to the amplifier should be considered. These are significant in valve- mounted amplifiers or mobile applications. If vibration or shock is present, levels must be determined and specified.

3.3 Temperature Range

For the majority of applications an amplifier capable of working in an ambient temperature range of 0 to 50 degrees is normally satisfactory. This is within the temperature range of most commercial electronic components and consequently, gives the most cost effective amplifier. Greater temperature ranges are possible using military specification components but the user should check availability and that the higher cost is acceptable.

3.4 Cooling Requirements

If the amplifier is mounted inside a sealed enclosure there should be sufficient space for air to circulate and sufficient surface area on the inside of the enclosure for loss of heat through the wall of the enclosure. Alternatively, a fan may be needed to increase air velocity inside the enclosure.

The use of a pulse width modulated output signal to the valve will result in a significant reduction in heat generated by the output stage of the amplifier. Check power dissipation figures if given.

3.5 Power Supply Requirements

Few, if any, amplifiers for electrohydraulic valves work directly from AC mains, most require a DC power input and to simplify installation it is common for amplifiers to be designed to accept just one power input, typically 24 volts, although a few amplifiers may also require ±15 volts at low power. If using an existing 24 volt power supply, check that the voltage is maintained within a range suitable for the amplifier.

When calculating the total current requirements it may be possible to reduce the size of the power supply unit by allowing only for the number of electrohydraulic valves which may be energised simultaneously and allowing sufficient current for these valves. Remember that a valve which is not de-energised but is driven to its null position with a zero demand signal may still draw current.

The operation of the amplifier in pulse width modulated modes can cause ripple on the supply lines and a large capacitor close to the amplifier can help reduce this ripple. The use of large capacitors on the power supply can significantly increase the inrush current when switching on the system and allowance must be made for this in power supply design.

4 INSTALLATION GUIDELINES

To prevent any inconsistencies in proportional valve operation, there are certain guidelines to be adhered to with regard to the electrical installation.

If a Eurocard mounted amplifier is to be incorporated into a control panel, it must be adequately screened from mains supply cables and other electrical/electronic equipment, in particular power supplies, relays and mains supply cables.

When wiring the amplifier to the valve, consideration must be given to the installation of cables. Input signals, feedback sensor (if required) and solenoid lines need to be wired with a screened cable, leaving one end open and the other end connected to earth (usually at the control panel). Cables to the valve need to be installed away from power supply lines.

Figure 5.1 shows a typical installation for a proportional valve with two solenoids and a feedback sensor. Note also that the amplifier in the control panel is mounted away from power supplies and other electronic items.

Use of a valve with an integrated amplifier can simplify installation. In this case, a single multicore, screened cable is required from the control panel. Care must be taken to ensure that the signal cable is adequately screened and installed away from power supply lines.

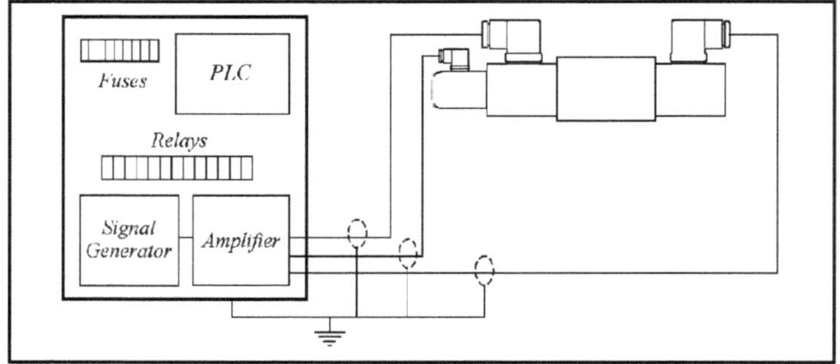

Figure 5.1: Valve Installation - 1

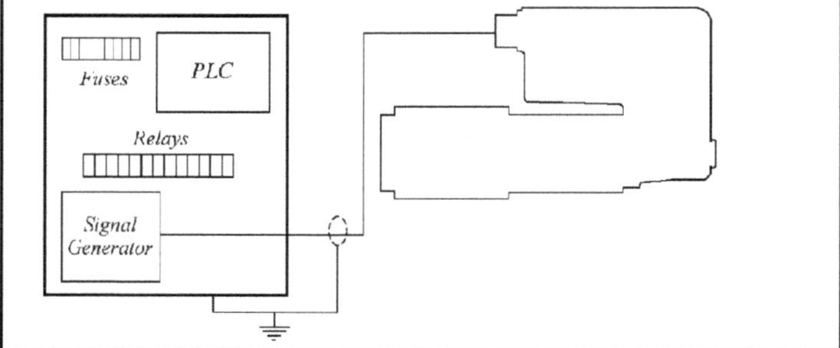

Figure 5.2: Valve Installation - 2

5 AMPLIFIER FEATURES AND FUNCTIONS

5.1 Input Signal Requirements

The drive signal for an electrohydraulic valve amplifier must be generated from a suitable source. Compatibility between signal and amplifier 0 volts should be checked. Large currents on long 0 Volt lines can cause changes in the input signal at the amplifier and a separate signal 0 Volt line is recommended. The signal 0V and power 0V must only be connected together at one point and ideally this point should be at the amplifier.

Typically, uni-directional valves, such as pressure controls, will use 0 to 10V input while bi-directional valves will use bi-direction input signals as ±10V.

During system design it should be established which direction of valve movement will result from +10V signal and the electronics for signal generation designed accordingly. If it is difficult to predict the required direction of movement then an amplifier which can invert the input signal may be useful.

It is always worth checking for high frequency voltage ripple on the input signal. High levels of ripple or noise can degrade amplifier performance causing unexpected level shifts and in extreme cases instability. This occurs if, for example, the circuits are forced into saturation at the peaks of the noise resulting in asymmetrical clipping and hence a shift in the average voltage.

5.2 Input Potentiometers

Where the input signal comes from an external potentiometer it may be useful to supply power to the potentiometer from the amplifier. Some amplifiers provide sufficient power at, for example, ±10V to allow one or more potentiometers to be connected directly to the amplifier. Care must be taken to ensure that the allowable output current from the amplifier is not exceeded when potentiometers are connected to the amplifier.

5.3 *Current Inputs*

Current inputs are used as an alternative to voltage inputs and are advantageous in systems operation in environments with high electrical noise or a long distance between parts of the system. The signal which must come from a suitable current source is less likely to pick up interference and is invariant to changes in cable or connector resistance. The range of 4 to 20 mA is normal.

5.4 *Switched Preset Inputs*

In some applications satisfactory operation may be achieved by switching between a number of preset inputs. Amplifiers with this feature usually switch between these inputs using on/off signals from a suitable source such as a Programmable Logic Controller (PLC). Care must be taken when interfacing PLC output signals to the preset input terminals on the amplifier card. Preset signals are normally fed from the cards own internal supply before being wired into the volt-free contacts of the PLC. If the PLC outputs have been configured to operate at different voltages, an interface relay must be used. The PLC output will energise this relay. The amplifier should then be wired into a volt-free contact on this relay, thus interfacing the different voltage levels.

5.5 *Input Signal Scaling*

Input signal scaling allows full valve output to be obtained with a user specified voltage. This scaling may allow between 5 and 10 volts to give full output. This is sometimes called gain adjustment. Alternatively, input scaling may allow 10 volts to give less than full output.

5.6 *Null Adjustment (Offset)*

To ensure that a zero input signal puts the valve precisely at its hydraulics null, zero offset adjustment is usually provided on the amplifier. When using pressure controls, a shift in the null will give either an output pressure with zero input signal or conversely will maintain pressure at zero until the input signal level is above the offset setting.

5.7 *Ramps*

Ramps are used to limit system shocks by limiting the rate of change of the signal applied to the valve. Ramps can significantly improve the feel of a manually-operated system by eliminating shocks during load reversal. An amplifier may have one, two or four ramps which are selected automatically depending on whether the input signal is demanding an acceleration or deceleration of the driven actuator and/or on the direction of amplifier movement relative to null. Such ramps used in conjunction with switched preset levels can form a complete movement profile with a minimum of external circuitry.

5.8 Ramp Enable/Disable

Some amplifiers allow ramps to be disabled by a separate input signal, this provides the option of rapid valve movement during part of the cycle, if desired.

5.9 Deadband Compensation

Some electrohydraulic valves are made with an overlap zone or deadband at null to reduce leakage and/or give a well-defined hydraulic condition when de-energised. This deadband can be eliminated from the apparent steady state performance by a compensation circuit that causes the spool to jump to the end of the overlap region. The magnitude of jump can usually be adjusted to match the overlap in the valve. In some amplifiers two deadband adjustments are provided to allow for different overlap each side of the null.

When the demand signal to the amplifier is zero, the spool should be at the centre of its null zone. For this reason, the deadband jump circuits are non-operative with the input signals less than a preset level. When the input signal exceeds this level the deadband circuit operates to move the spool to the end of the overlap region.

5.10 Signal Monitor Points

Commissioning and fault finding are made easier if signals in the amplifier can be monitored. The ability to monitor the position or pressure demand signal after the application of ramps and deadband jumps is very useful. This allows ramp times to be set to a sufficient length to avoid shocks before enabling the valve. With open-loop valves it is also useful to be able to monitor solenoid currents. For closed-loop valves it is usually more important to monitor the spool position.

5.11 Output Enable/Disable (Inhibit)

An output enable/disable function is used to de-energise the valve while keeping the amplifier operative. This allows for adjustments to be made to functions such as ramp time and deadband using the signal monitor point as described above.

This feature can sometimes be used in combination with interrupting the power supply to the amplifier under emergency stop conditions to speed-up movement to the valve failsafe condition. In systems with large inertia, sudden de-energisation of the electrohydraulic valve may give an unacceptable system shock and some hydraulic means of limiting shock should be considered.

5.12 Detection of Loss of Feedback Signals

Electrohydraulic valves with spool position or pressure feedback will fail to operate correctly if feedback signals fail for any reason. It is always advisable to use an amplifier which disables the output if errors in the feedback signals are detected. If a fault is detected it may be desirable for the amplifier to remain disabled until a reset signal is received or to restart immediately the fault condition is removed. Few amplifiers offer both options.

5.13 Current Limiting (Short Circuit Protection)

To prevent damage to the electrohydraulic valve, the amplifier output should ideally be short circuit protected. If a short circuit is detected it may be desirable for the amplifier to remain disabled until a reset signal is received or to restart immediately the short circuit is removed. If this is important, check the operation of the amplifier you plan to use.

5.14 Error Output Signal

As the valve and amplifier form an important part of a system, it may be desirable for the complete machine to shut down if the amplifier detects a sufficiently large error. For this reason the generation by the amplifier of a "No Error" signal would be useful. Such signals could be monitored by the system controller which would be programmed to take appropriate action if the "No Error" signal was lost.

5.15 LED Fault Indicators

To assist with fault finding, it is helpful to have a number of LED's to indicate normal or error conditions. These can include presence of power supply, ramp enabled, output drive enabled, power supply overload and LVDT fail, etc.

6 SYSTEM DESIGN AND CONSTRUCTION

During the design and construction of any system we should give due regard to all of the potential risks and hazards including those associated with any failure of the control system.

Machine guards and operation on loss of electrical power are the obvious items to consider. However, there are a number of safety and quality standards that have to be observed and it is recommended that the design review is formal and when necessary formal tests conducted.

6.1 Standards and Directives

To help harmonise safety and quality standards within the EU a wide range of Directives and harmonised standards have been generated. These harmonised standards describe design and operational requirements that are aimed at maintaining the health and safety of the user. The standards are applicable to any product that is intended for use within the EU.

Fortunately, other countries, such as the U.S., have worked with European Standards Organisations so that U.S. and European requirements are very similar.

The European standards are grouped according to applications rather than by the technology used in the construction of the product. Consequently, an electrohydraulic system may come under one or more standards. In Europe the manufacturer of the product or system is legally obliged to ensure that the product conforms to all applicable standards.

A list of the applicable Directives and standards is published by the EU and can be found in the EU web site at: http://europa.eu.int/comm. The obvious Directives to consider are the Machinery Directive and the EMC Directive. Details of the EMC Directive can be found at http://europa.eu.int/comm/enterprise/electr_equipment/emc/index.htm. This site lists the currently active standards applicable to electrical equipment. The list is the same as that published in the "Official Journal". For more information on affixing the CE logo and on CE Marking see BFPA/P61 *'The Safety of Machinery Directive – A Guide to the Use of CE Marking'*.

Some of the standards applicable to electrohydraulic systems are listed and discussed below.

6.2 Temperature, Humidity and Vibration

The most comprehensive current standard for testing industrial equipment for susceptibility to these three environmental conditions is IEC 68. A wide range of test severity levels are given in this standard and consequently the levels used for the unit under test must be specified.

Equipment that is subject to a combination of high humidity and temperature cycles should be tested to IEC 68 to check that water vapour is not drawn in through the case.

Such tests are additional to IP ratings that indicate the amount of ingress protection with regard to solids and liquids.

6.3 Electromagnetic Compatibility

All products put into use within the EC must conform to the EEC Directives 89/336/EEC, 92/31/EEC and 98/31/EC on electromagnetic compatibility.

These Directives are not technical documents and the main product performance requirements are detailed in BS EN 50081 and BS EN 50082. These are supported by BS EN 61000 EMC.

6.4 Emissions

The maximum electromagnetic noise emitted by electronic equipment is subject to limits that specified in BS EN 50081-2.

6.5 Immunity

Products are required to pass the immunity tests conducted at the levels detailed in BS EN 61000-6-2:2001.

6.6 Power Supply Faults

If necessary, the equipment may be tested for susceptibility to variations in supply voltage or short interruptions. The applicable test method for low voltage power supplies are given in BS EN 61000-3-3:1995.

7 REFERENCES

European Directives

73/23/EEC*	Low Voltage
87/404/EEC*, 90/488/EC	Simple Pressure Vessels
89/106/EEC*	Construction products
89/336/EEC*, 92/31/EEC, 98/13/EC	Electromagnetic compatibility (EMC)
98/37/EC	Machinery
94/25/EC	Recreational craft
94/62/EC	Packaging and packaging waste
95/16/EC	Lifts
98/85/EC, 96/98/EC	Marine equipment
97/23/EC	Pressure equipment
2000/9/EC	Cableway installations designed to carry persons
93/68/EEC	Council Directive amending Directives marked with *

European Technical Standards

BS EN 50081-2:1994		Generic Emission Standard for Industrial Equipment Industrial
BS EN 61000-3-2:2001		Limits on Harmonic Current Emissions
BS EN 61000-3-3:1995		Limits in output voltage fluctuations for power supplies
BS EN 61000-4-1		Testing and Measurement Techniques
BS EN 61000-4-2		Electrostatic discharge
BS EN 61000-4-3		Radiated Electromagnetic energy
BS EN 61000-4-4		Fast transients
BS EN 61000-4-5		Surge immunity
BS EN 61000-4-6		Conducted interference
BS EN 61000-4-11		
BS EN 61000-6-2:2001		Generic Immunity Standard for Industrial Equipment. Supersedes EN 50082-2
IEC 68	2-1 Ab, Ad	Low temperature storage and start-up
	2-2 Bb, Bd	Dry heat storage and operation
	2-3 Ca	Damp heat operation
	2-6 Fc	Sinusoidal vibration
	2-14 N	Thermal shock
	2-30 Db	Damp heat cycle
	2-38 Z/AD	Composite temperature and humidity
IEC 60529		IP rating method. Also see BS EN 60529

Other Standards Applicable to Specific Application Area

Control Devices and Proximity Sensors
BS EN 50227:1999

Lifts and Escalators
BS EN 12015:1998 Emissions
BS EN 12016:1998 Immunity

Low Voltage Switchgear
BS EN 60439-1:1999
BS EN 60947 parts 1 to 6

Medical Equipment
BS EN 60601-1-2-2002

Programmable Controllers
BS EN 61131-2-1995

Safety of Machinery

BS EN 982	Safety of Machinery - Safety requirements for fluid power systems and their components - Hydraulics

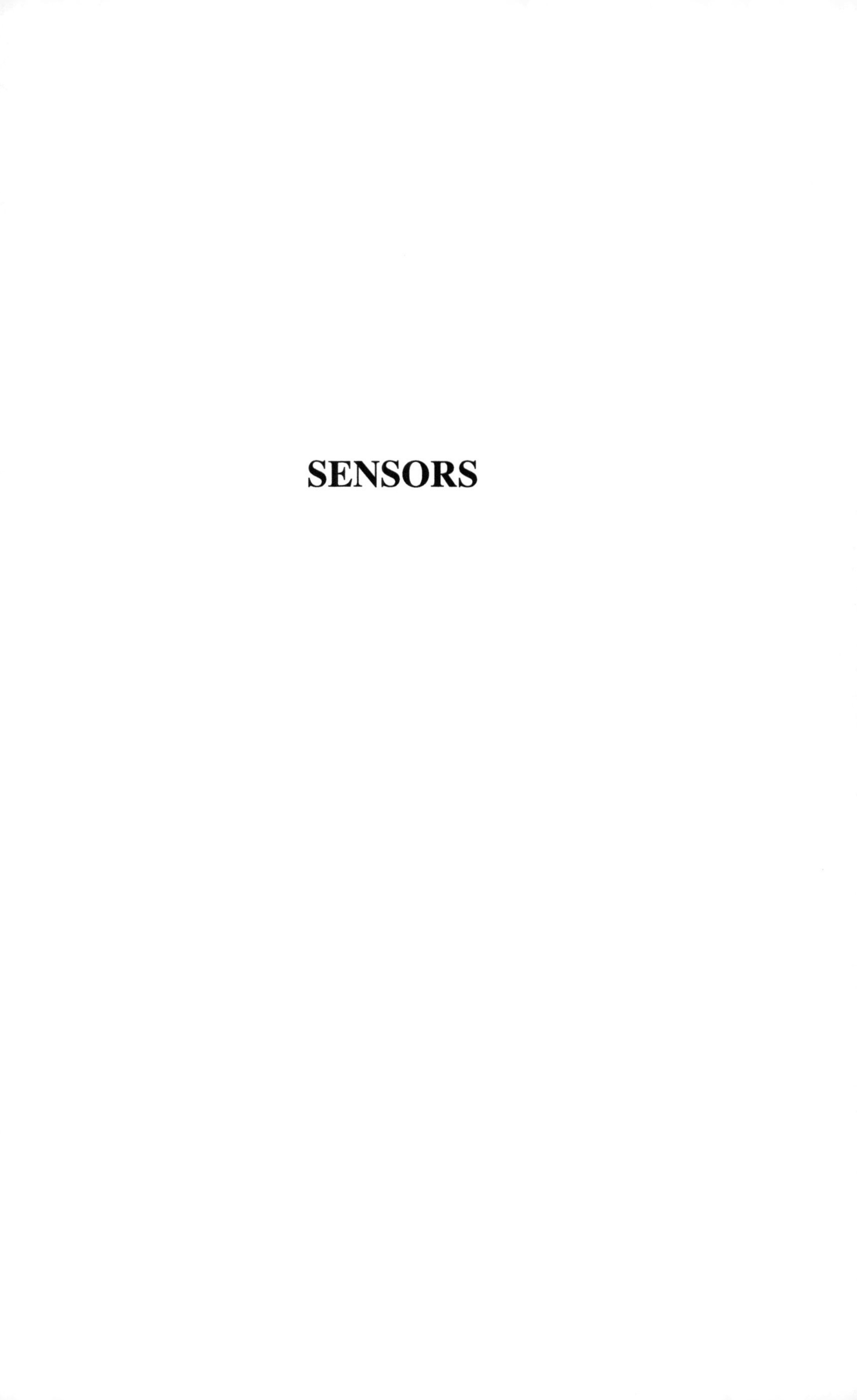

SENSORS

1 GENERAL

The selection of a sensor for any application will depend on both the required electrical characteristics and the mechanical environment in which the sensor is to be used. When necessary the sensor must be adequately protected against aggressive environments whether these are chemical, mechanical, electrical or climatic.

Any sensor operating by induced or applied magnetic fields may be influenced by internal or surrounding metalwork, such as a pressure tube or, for example, a large flywheel in close proximity.

An environment with high levels of shock and vibration will require special mounting considerations to attenuate shock while maintaining sufficient signal accuracy.

In general the stroke of the mechanical sensor should exceed the maximum stroke of the actuator to prevent unintentional damage to the sensor. When mounting a linear sensor onto an actuator the alignment between the sensor and actuator should always be checked as misalignment may cause mechanical damage or inaccuracy. Mounting the sensor inside the hydraulic cylinder is beneficial with respect to both alignment and providing protection from the environment.

If the sensor output requires processing by an electronic circuit to obtain a usable output signal an excessive distance between the sensor and the electronics may degrade performance. It is not always possible to filter out noise caused by any form of interference as all filtering causes a phase lag in the signal and excessive lags will degrade system performance.

2 ANALOGUE POSITION SENSORS

For these guidelines an analogue position sensor is defined as a sensor that converts a continuously variable position of the measurand into a continuously variable voltage or current signal.

2.1 Potentiometers

Potentiometers are typically low cost and widely available. Phase lag can be negligible provided the mechanical connection to the potentiometer is sufficiently rigid.

Potentiometers are simple to use as they require only a DC voltage and little or no additional circuitry. The potentiometer is a contacting device and so has a finite life.

Consequently, potentiometers are suitable for applications having moderate use or where the position sensor can be easily replaced.

Potentiometers are available in rotary and linear forms. Rotary potentiometers are available as single turn - typically limited to about 320°, or multi turn - typically 3 or 10 complete turns.

For position sensing a potentiometer requires at least three electrical terminals; one at each end of the resistance element and one to the wiper. In use a steady DC voltage is applied across the resistance element. The output voltage, taken from the wiper terminal, depends on the position of the wiper relative to the two ends of the resistance element. The ends of the resistive element may be supplied with equal positive and negative voltages so that the centre position will give zero output. However, the output voltage will be modified by the load resistance. To limit loss of linearity the load resistance should be high relative to the potentiometer resistance (see figure 6.1). This error can be reduced by the use of a centre tapped potentiometer which also minimises drift around this position.

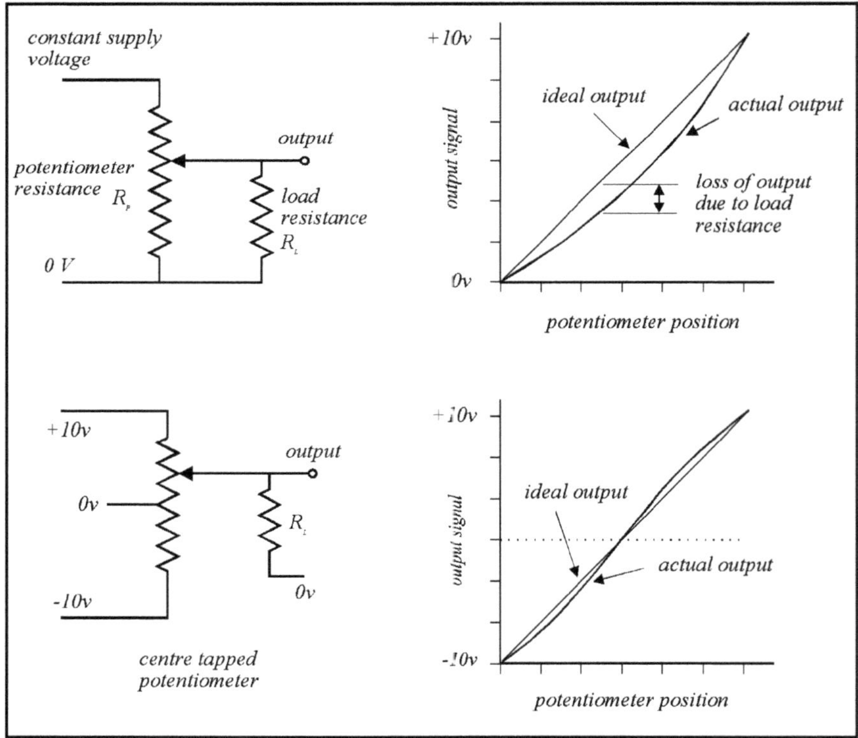

Figure 6.1: Potentiometer Linearity

The output signal from a potentiometer is less susceptible to noise if the resistance of the potentiometer is low but the lower the resistance, the greater the current taken by the potentiometer, which may lead to overheating.

There are three major potentiometer constructions, wire wound, conductive plastic and hybrid. The wire wound potentiometer has limited resolution, whereas the conductive plastic and the hybrid can have near to infinite resolution.

If the potentiometer is subject to high shock and vibration, the spring load on the wiper must be sufficient to maintain contact with the track. However, excessive loads can reduce the life of the potentiometer.

Continuous reversal of the wiper at one point on the resistance element will cause local wear.

Some potentiometers can be oil immersed allowing them to be mounted, for example, inside a cylinder. In this case the signal wires need to pass through a pressure-proof connector.

Other application considerations include linearity, temperature drift, hysteresis or deadband caused by the mechanical connection between shaft and wiper, shaft run out or end play.

2.2 Linear Variable Differential Transformer (LVDT)

2.2.1 General

The LVDT is an electrically non-contacting transducer used to give an analogue output proportional to linear position. The LVDT works on the principle of magnetic induction.

2.2.2 Construction

The simple sensor consists of a moveable iron core surrounded by three coils (a primary and two secondary coils). The centre primary coil is supplied with an AC excitation voltage, typically of between 5 and 10 kHz. The magnetic field produced by the primary induces voltages in the two secondary coils. The relative magnitude of the two secondary voltages is dependant on the position of the core. When the core is central to the two secondary coils there is no difference in voltage and no net output (see figure 6.2).

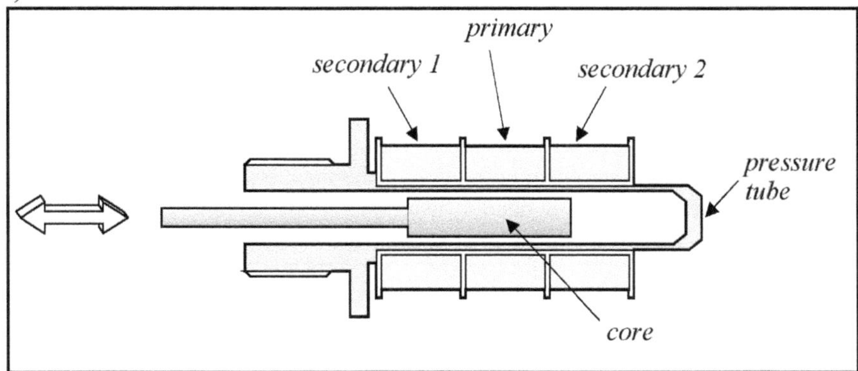

Figure 6.2: LVDT Construction

Subtracting the induced secondary output voltages one from the other produces an AC signal the amplitude of which is proportional to the position of the core. Further electrical processing is required to convert this signal into a steady DC output and detect the difference between positive and negative travel. For ease of use some LVDT's have the required electronics built into the LVDT housing.

LVDT's have very high resolution, low hysteresis and high reliability but can suffer from temperature drift. For fluid power applications the LVDT has the advantage that it will still operate when the core is oil immersed and separated from the coils by a non-magnetic pressure tube.

As the core is the only moving part the rest of the LVDT can be potted, or hermetically sealed, so that the sensor can work in damp or corrosive atmospheres.

Whilst there is no contact at the magnetic sensing element, bearings or guides are usually needed to maintain a constant clearance between the moving armature and the pressure tube.

A disadvantage of LVDT's is that the de-modulation process never eliminates all noise from the output signal and excessive de-modulation noise makes sensing velocity by electrical differentiation difficult. Operating the LVDT at a higher frequency can make it easier to filter out the noise with less phase lag but high frequency operation will increase the temperature drift of the sensor. The stability and performance of the LVDT depends on all the working elements including the core material which should be treated with care.

2.3 Eddy Current Devices

Eddy current devices are similar in construction to inductive devices but operate at much higher frequencies. At these higher frequencies the conductivity of the core or target material is more important than the magnetic properties. In fact, non-magnetic cores or targets can be used.

Electronics are required to measure the change in inductance in that coil as the core material or concentric sleeve moves. Lower temperature drift is possible if the electronics measures the change in the resistive rather that the inductive vector in the coil.

The eddy current sensor is an electrically non-contacting device. Some sensors may be mounted inside a cylinder without any physical extension to the length of the cylinder.

2.4 Hall Effect Sensors

Hall Effect sensors are based on semiconductor devices that produce an electrical voltage proportional to the strength and direction of the magnetic field passing through it. The magnetic field is usually produced by a permanent magnet and the output from the Hall Effect sensor changes as the magnet moves relative to the sensor.

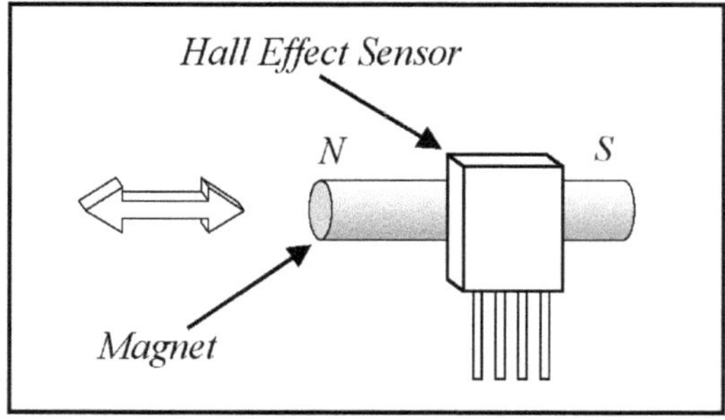

Figure 6.3: Hall Effect Sensor

To produce a linear position signal the magnet is usually moved relative to the sensor shown in figure 6.3. However, linear output is only obtained over the middle 1/3 of magnet length. This arrangement allows the magnet to be mounted inside a non-magnetic pressure tube with the Hall Effect sensor on the outside.

The Hall Effect sensors can provide very high bandwidth with almost negligible phase lag.

2.5 Magnetostrictive (Strain Wave) Position Sensors

These position sensors are used for linear position sensing often being located inside the cylinder. These sensors consist of a tube of magnetostrictive material, as shown in figure 6.4, used to convert a current into a torsional pulse or strain wave. The effective length of the tube is modified by a magnet which moves axially along and outside of

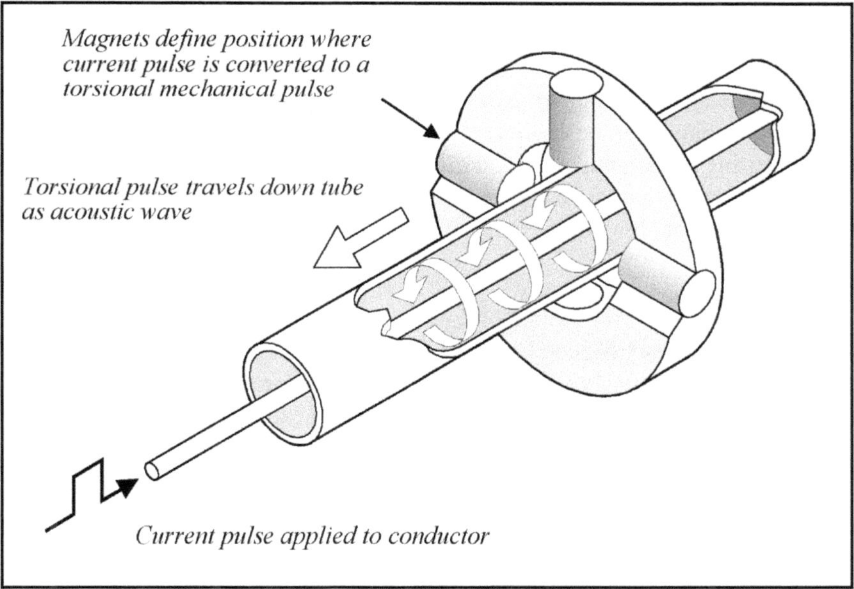

Figure 6.4: Magnetostrictive Sensor

the tube. When the sensor is used inside a hydraulic cylinder the magnet is typically attached to the piston. Position information is derived by measuring the time taken for the torsional pulse to travel down the tube. The time delay inherent in the sensor due to transit time of the strain wave can, particularly at long strokes, introduce noticeable phase lags. The phase lag due to strain wave transit time is often significantly increased as it is often necessary to average a number of consecutive readings to obtain good resolution.

As no contact is required between the magnet and the magnetostrictive waveguide assembly there is no mechanical wear to reduce life expectancy.

2.6 Capacitive Sensors

Capacitive position sensors operate by detecting the change of the capacitance resulting from the relative movement of two electrodes or plates. These can also be one cylinder moving inside another consequently the mechanical construction can be robust.

These sensors can be very accurate over short strokes but at large strokes accuracy and phase lag are poor.

3 ANALOGUE VELOCITY SENSORS

3.1 Tachogenerators

Tachogenerators are used for measuring rotational velocities.

In construction they are similar to a small electric motor. The linearity can be better than 0.1% but commutator ripple needs filtering particularly for low speed operation. Three-phase tachogenerators have less ripple and are consequently easier to filter.

Tachogenerators may be used within a position control loop to provide "velocity damping" which allows a higher gain in the position loop.

Tachogenerators can be connected directly to the output shaft eliminating backlash and gearing but alignment is important to avoid load on bearings. They have no offset errors (zero output at zero speed), low temperature drift, excellent linearity, bi-directional operation, low inertia and low load torque.

3.2 Induction Tachometers

These and other AC tachogenerators have the advantage of no commutator and consequently higher reliability. In addition, linearity and temperature drift can be better than DC tachogenerators. However, the induction tachogenerators may have a null bias which results in a zero output at a non-zero speed.

4 ACCELERATION SENSORS

The typical construction of an acceleration sensor is to have a reference mass supported by a flexible member. Deflection of the mass is proportional to the acceleration. Accuracy and linearity of the sensor depend on the perf-ormance of the deflection measuring sensor. This sensor may be one of the position sensors described above. The stiffness of the sensor and mass of the reference load defines the natural frequency which needs to be higher than the bandwidth of the hydraulic system. To obtain high frequency operation the flexible member must be stiff which results in small amplitude movement and reduced accuracy.

5 FORCE SENSORS

5.1 Load Cells

If closed-loop control of force is required this may be achieved by measuring cylinder pressures. In some systems this may be significantly easier to implement but for high accuracy a load cell may be advantageous.

6 PRESSURE TRANSDUCERS

Pressure transducers typically use a diaphragm which is deflected by the pressure. This deflection is then measured by one of the means discussed above. A simple pressure transducer may use a piston and measure its deflection using a linear potentiometer. Such devices suffer from leakage and low frequency response. The more commonly used types consist of a diaphragm with limited deflection and either strain gauge, capacitive or inductive position sensor.

A typical capacitive pressure transducer is shown in figure 6.5. The diaphragm of

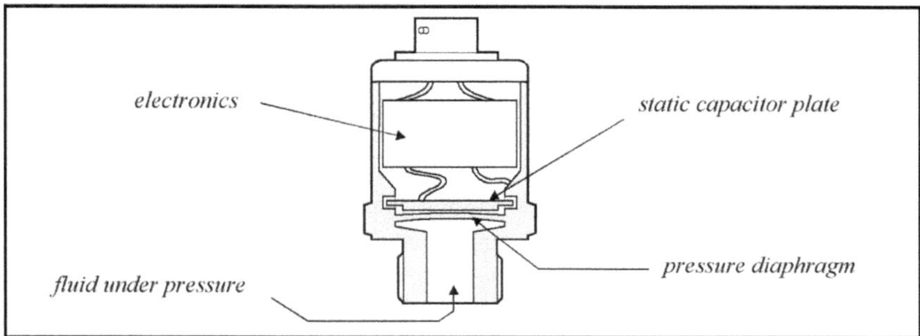

Figure 6.5: Capacitive Pressure Transducer

the sensor is used to form one of the plates while the other plate is placed very close to the diaphragm to produce large variations in capacitance for small movements.

If fluid is supplied to both sides of the diaphragm a differential pressure can be measured. Such transducers can suffer from a reduction in sensitivity with increased mean pressure and may exhibit hysteresis on pressure reversal.

As an alternative to a differential pressure transducer, two independent transducers can be used and electrical means used to obtain a difference signal.

7 DIGITAL POSITION SENSORS

7.1 Definition

The digital position sensors described below all operate by dividing the movement of the measurand into discrete steps. When the measurand moves a distance equivalent to

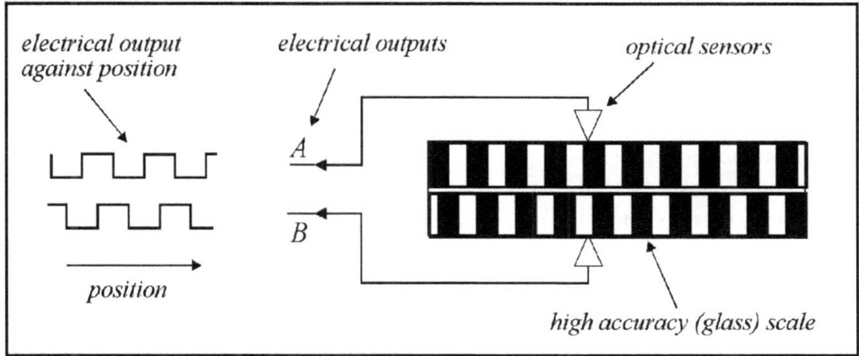

Figure 6.6: Digital Position Sensor

one step the output changes by one digital bit or state. Between these steps the position of the measurand is unknown and cannot be inferred or calculated.

In operation, digital electronics may be necessary to count the steps and/or calculate the position. The steps are usually supplied by a high precision scale in either glass or a magnetic material. The pattern or steps may be printed, etched or in the case of a magnetic strip recorded onto the scale. The scale may be part of the actuator or cylinder rod or part of a machine, for example, the teeth on a gear. See figure 6.6. The teeth on the cylinder shaft must be covered by a non-magnetic material such as chrome or a ceramic.

7.2 Incremental Encoders

Incremental encoders are available for measuring either linear or rotary displacement. The resolution of rotary encoders is defined by the number of pulses per revolution, or for a linear encoder by the number of pulses per millimetre.

These encoders can only measure relative displacement, consequently, if an absolute position output is required the actuator must first move to a known position. Two separate signals are necessary if the sensor is to measure both magnitude and direction of movement. The advantages are high resolution and low temperature drift.

8 ABSOLUTE POSITION ENCODERS

The mechanical construction of absolute encoders is very similar to incremental encoders, the difference being an increased number of parallel outputs. The outputs can switch independently between on and off states and the pattern of the output states provides a code that represents absolute position. The pattern of output states of most absolute encoders conforms to the Grey code in which only one output changes at a time. This eliminates the errors that are possible using a binary system in which more than one output can change between valid output states.

8.1 Advantages and Disadvantages

The advantage of an absolute encoder is that position can be read directly on power up without requiring the load to be moved to the datum position. The disadvantage of absolute encoders is the number of outputs required in high accuracy systems.

9 DIGITAL VELOCITY SENSORS

9.1 Digital Tachogenerators

These work on a similar principle to incremental encoders and may use optical, inductive or magnetic sensing.

To obtain a velocity signal only one row of segments is necessary and additional electronics are used to detect the rate of passing over the segments. If a negative velocity signal is required when the shaft reverses, two outputs are necessary as in a conventional incremental encoder.

If the rotational or linear speed is high enough, the pulse rate can be measured and used as a speed indicator. At low speed the time to change one position count will give a better measurement of speed.

PRESSURE CONTROLS (TYPES)

1 PRESSURE CONTROL

In using a force controlled solenoid in a pressure control valve, the solenoid armature is used to react directly against a force produced by the system pressure. A cross-section of a representative pressure control valve is shown in figure 7.1. The solenoid armature is connected to a moving valve poppet which can be pushed into a small cylindrical hole to reduce, or even completely close a flow path. This establishes the variable area in this type of valve. As the area is reduced, so the pressure drop through the valve is increased and since the downstream chamber is connected directly to the system reservoir, the pressure in the upstream chamber is increased. This pressure acts on the valve poppet and so produces an opposing pressure dependent force to react against the electromagnetic force of the solenoid.

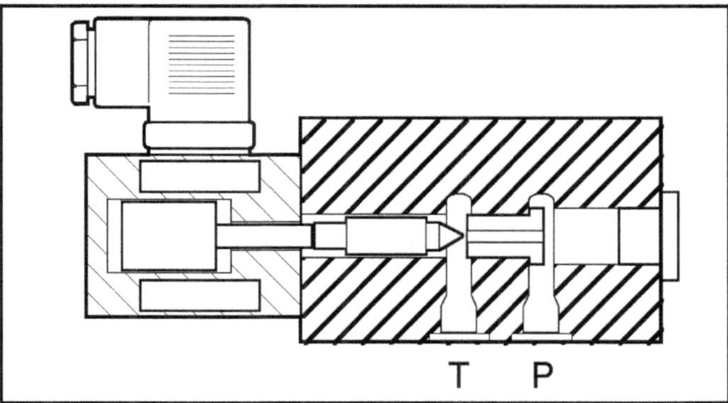

Figure 7.1: Pressure Force Controlled

Valves of this design are not able to pass large flows because of the limited force available from the solenoid. However, they can easily be used to act as part of a two-stage assembly, very like a conventional relief valve, as shown in figure 7.2. Hence the solenoid stage acts as a pilot valve to control the pressure acting on the rear face of a larger, two-port, poppet valve. The force balance across this second poppet establishes an upstream pressure which tracks any changes in the pilot pressure. This design can be used to control pressure in systems capable of passing very high flows.

Stroke controlled proportional solenoids may also be used in pressure control designs, as shown in figure 7.3. This again includes a displacement transducer but in this

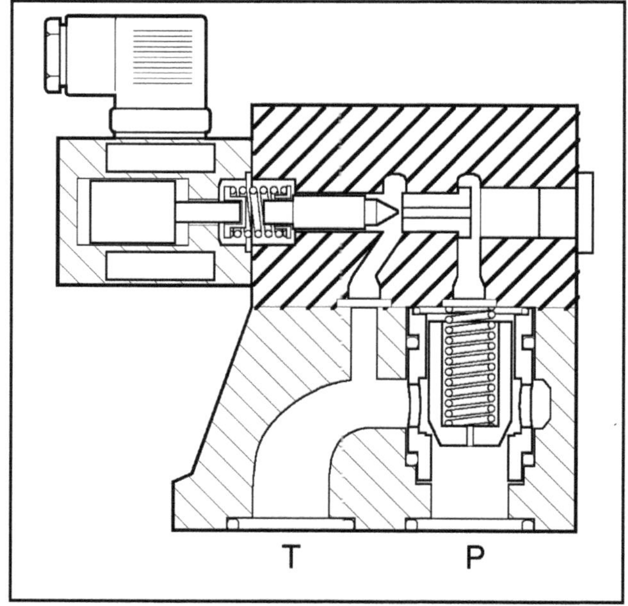

Figure 7.2: Relief Valve Assembly

case it is the position of one end of a spring which is being controlled in a feedback loop. The solenoid armature is used to move this end of the spring and provides the force necessary to compress the spring. The other end of the spring is attached to a poppet control element as described above. In fact, this design is much closer to a conventional manually set relief valve but with the spring pre-compression established by the solenoid. This valve can also be used to pilot operate a larger poppet and control a system pressure.

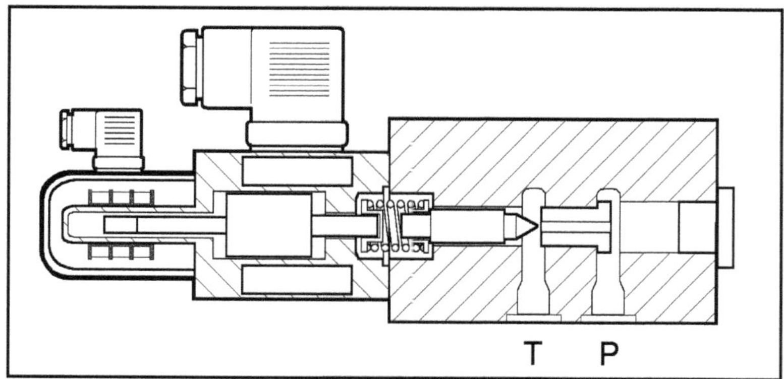

Figure 7.3: Pressure Pilot, Stroke Controlled Solenoid

Pressure reducing valves can also be pilot operated by either force or stroke controlled solenoids. This will then allow the pressure in part of a system to be controlled from an independent electrical signal at a reduced level. There are two basic types of reducing valve and these give rise to the different operating characteristics described in a later section. In the design referred to as type B and shown in figure 7.4, the flow to the pilot stage is taken from the upstream pressure point via the constant flow valve shown. The pressure set by the pilot acts on the upper surface of the main valve element and balances the reduced pressure at the level set by the pilot valve. A small non-return valve is usually built into the design and is shown here within the main valve element to allow some return of flow from the reduced port. The type A design takes pilot flow from the reduced port through a fixed orifice. The pressure balance across the main valve is maintained as in the design of figure 7.4.

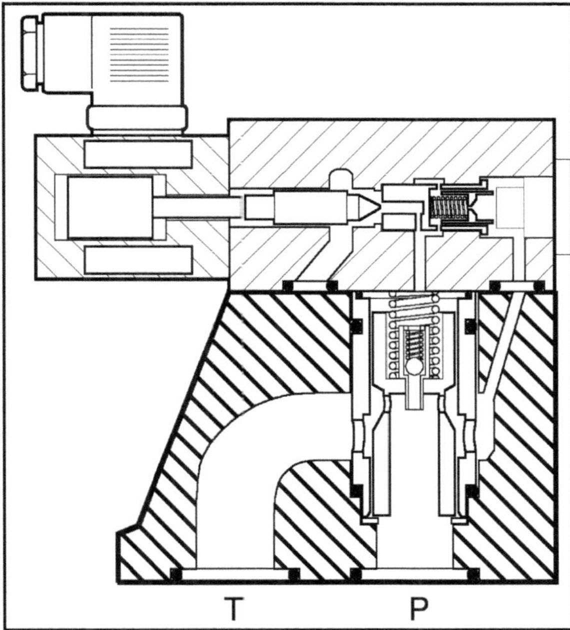

Figure 7.4: Pressure Reducing Valve

SELECTION OF PRESSURE CONTROLS

1 SCOPE

Electrohydraulic pressure controls are normally used when it is required that the pressure in some part of a hydraulic circuit be held at or below a level which is to be set and controlled by an electrical signal.

An electrohydraulic valve makes it possible to change the pressure during the machine cycle to follow system pressure requirements, minimise energy consumption or protect weaker parts of the hydraulic circuit from excess pressure.

The description of performance characteristics in this section is limited to pressure relief and reducing valves. Although there are other types of pressure control they are not normally available in electrically modulated form.

It is important to remember that relief and reducing valves have fundamentally different internal operation and cannot be functionally interchanged with each other.

1.1 Relief Valve Operation and Use

A relief valve is designed to open if the system pressure at its inlet port exceeds its pressure setting. The valve will divert just sufficient flow to tank to maintain the pressure and will close if the pressure drops below the set level.

A relief valve is typically used to control or limit the pressure in a system which is supplied directly from a pump or other prime mover.

1.2 Reducing Valve Operation and Use

A reducing valve is used to control the pressure in part of a hydraulic system to a level which is less than the main system pressure. This allows the use of a lower pressure for a particular service without installing a separate pump.

A reducing valve achieves this function by controlling the flow through the valve and will close to reduce the flow if the pressure at the outlet port approaches or exceeds the pressure setting.

2 PERFORMANCE CHARACTERISTICS – INTRODUCTION

The following description of performance characteristics is based, where possible, on BSI and ISO standards and reference to these standards is recommended if further information on test methods is required.

The most applicable standard for testing for performance characteristics is ISO 10770-3 - Hydraulic Fluid Power – Electrically modulated hydraulic control valves – Part 3: Test methods for pressure control valves, however, this is still under development. For methods of measuring the head loss through the valve see ISO 4411 or ISO 6403.

The discussion of characteristics is principally intended to assist with the selection and application of electrohydraulic pressure controls but some remarks will equally apply to mechanically set valves. In addition, some comments about performance can apply to both relief and reducing valves and where this is the case, they are referred to collectively as pressure controls.

The performance of pressure controls varies widely and particular mechanical constructions can provide special features which enhance some aspects of performance at the expense of others. Consequently the important system requirements should be identified prior to selection of the valve.

Attention is drawn to the two different types of reducing valve described in Section 7. In view of the significant differences in performance characteristics, they are often discussed separately and are hereafter referred to as Type A for valves taking pilot flow from reduced pressure and Type B for valves taking pilot flow from the high pressure inlet port usually via a flow control.

3 STEADY STATE PERFORMANCE CHARACTERISTICS

3.1 Maximum Flow

The maximum flow rate for a single-stage electrohydraulic pressure control will normally be about 4 L/min. For larger flows a two-stage valve will be necessary. An indication of typical flow rates possible with various valve interface sizes is given in Table 8.1.

SIZE (ISO)	3	5	6	8	10
TYPICAL MAXIMUM FLOW (L/M)	40	100	200	400	600

Table 8.1: Typical Maximum Flow, Two-Stage Electrohydraulic Pressure Controls

It should always be remembered that the peak flow seen by a pressure control will be greater than the steady state system flow. When changing pressure in a system, additional flow is required to compress and decompress the system. If rapid response is required and system volumes are large, the transient flow can be much greater than pump flow.

In use, single-stage valves may exhibit a very rapid increase in head loss beyond the recommended maximum flow due to saturating or high flow force effects. This can give rise to longer response times or high overshoots when subject to transient flows.

It is worth remembering that when operating at low-pressure levels the maximum flow through a system may be determined by the resistance in the pipework rather than the performance of the valve. If the pipework head loss is too great the system may not be able to achieve maximum flow and minimum pressure simultaneously.

Two-stage reducing valves of Type A will always have a flow limit beyond which they cannot operate. This limit occurs when flow forces on the main spool equal or exceed the spring force. Increasing the spring force will increase the maximum flow but at the expense of minimum pressure. This limit does not occur with Type B reducing valves.

3.2 Minimum Flow

All pressure controls must be supplied with sufficient flow to overcome leakage before they can function correctly and control pressure. Two-stage pressure controls will, while controlling pressure, require additional flow to meet pilot flow requirements which may be from 0.5 to 4 L/min.

A two-stage relief valve may, due to interaction between pilot and mainstage, exaggerate the leakage flow, increasing the minimum flow needed for correct operation. For large relief valves, up to 8 L/min may be required before full operating pressure is reached.

Most reducing valves can be expected to operate correctly down to zero flow.

3.3 Reverse Flow (Reducing Valves Only)

Two types of reverse flow are possible with each performing its own specific function. A reducing valve may provide one, both or neither of these functions depending on mechanical construction.

3.3.1 Reverse Flow to Inlet

When the reducing valve is applied between a directional valve and actuator, it will be necessary to reverse the flow through the reducing valve if actuator reversal is required. This reverse flow normally by passes the valve operating elements and will usually consist of a simple, low preload, check valve built into the valve body. This provides the minimum resistance to reverse flow. To open the check valve it is necessary for the inlet pressure to be lower than the pressure at the reduced pressure port.

3.3.2 Reverse Flow to Tank

An alternative form of reverse flow is required when the reduced pressure is expected to remain constant regardless of flow direction. Under such circumstances additional control lands are required on the main metering spool to discharge to tank the flow entering the reduced pressure port. Single-stage valves will normally have such a spool and reverse flow up to 20 L/min is common.

Two-stage reducer valves do not usually provide reverse flow to tank via the mainspool so reverse flow is then limited to low flows through the pilot, often only 1 or 2 L/min. To achieve high reverse flow capability the mainstage must be specifically designed with a direct connection to tank which bypasses the pilot.

3.4 Maximum Pressure

Clearly, a pressure control must never be used above its pressure rating defined by the manufacturer or supplier (see Section 8.5).

It should be remembered that in use the followed pressure level will vary with flow rate and any maximum pressure will only be obtained exactly at one particular flow (see figure 8.1).

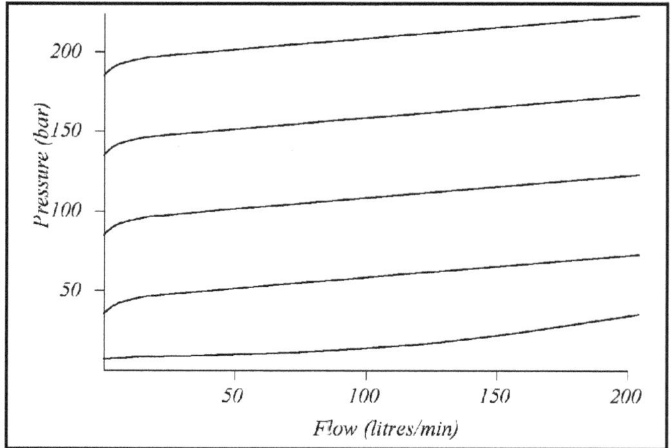

Figure 8.1: Typical Pressure/Flow Characteristics of a Relief Valve

Typical values of maximum pressure are 70, 150, 250 and 350 bar and due to characteristics stated above will only be achieved at one particular flow and some valve to valve scatter is unavoidable.

If using a relief valve at low flows it is worth checking that the desired pressure can be achieved with the flow available as the valve may have been set to give maximum pressure at a high flow rate. Conversely, if the maximum pressure was set at a low flow then greater than anticipated pressure will occur at full setting and with higher flow rates.

The magnitude of the pressure change is highly dependent on the design of the valve but typically a reducing valve may vary only 2 to 5 bar, a two-stage relief valve may vary 10 to 30 bar, while a single-stage relief valve may vary up to 50 bar across its flow range.

In a single-stage electrohydraulic pressure control the only method of limiting the maximum pressure is by limiting the maximum current. Current limiting circuits should be included in the drive amplifier design when required.

3.5 Minimum Pressure

The minimum pressure setting also known as the head loss is normally achieved with zero electrical signal, i.e. zero current. Minimum pressure varies with flow through the valve and may change considerably over the operating range.

When de-energised, single-stage relief valves normally have zero cracking pressure and approximate to a fixed orifice characteristic as flow increases. Two-stage relief valves are usually closed by the mainstage spring and will require a minimum pressure of between 0.5 and 5 bar to open the valve.

A typical value of minimum pressure at maximum flow rate is 8 bar but valves can range from 2 to 20 bar. Two-stage valves with a sprung-open mainstage will give lower minimum pressure but require a dedicated constant flow source to operate the valve and are not in common use.

Reducing valves of Type B may also be sprung-open or sprung-closed. The minimum reduced outlet pressure can be near zero for a sprung-closed valve while sprung-open valves will require 2 to 4 bar at the reduced pressure port to close the valve and control the pressure. Type A valves can only be sprung-open and will normally have a higher mainstage spring preload and may have a minimum pressure of up to 10 bar.

3.6 Pressure/Flow Characteristic (Relief Valve)

An increase in flow through a relief valve should always result in an increase in pressure. This characteristic is normally referred to as the override (figure 8.1).

It would seem reasonable to reduce the override to obtain a more consistent pressure over the flow range but a low override relief valve will have marginal stability and will be prone to resonance and pressure overshoots.

3.7 Pressure/Flow Characteristic (Reducing Valves)

Reducing valves will normally exhibit a decrease in pressure as the flow through the valve increases. This characteristic seen in figure 8.2 assumes a constant inlet pressure across the full flow range.

The magnitude of under ride is a function of valve design. A Type A reducing valve which takes pilot flow from the outlet pressure port can have a change in pressure with flows less than 1 bar over its full flow range but is dependant upon the quality of the electrohydraulic pilot valve.

A Type B reducing valve will normally have a very consistent under ride of between 4 and 8 bar over the full flow range.

A reducing valve has the advantage that it can have zero override without loss of stability and is an effective way of obtaining higher pressure consistency.

A pressure control characteristic often overlooked is the change in reduced pressure with changes in inlet pressure. An increase in inlet pressure will cause a small reduction in reduced pressure, typically 1/50th of the change in the inlet pressure.

Some designs of reducing valve show a step up in pressure as the flow at the reduced pressure port reaches zero. A single or two-stage valve designed to give re-

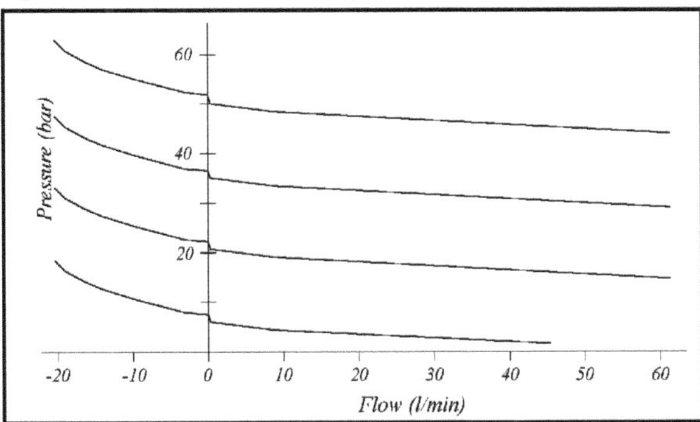

Figure 8.2: Typical Pressure/Flow Characteristics of Reducing Valve with Reverse Flow Capability

verse flow capability on the mainspool may be slightly overlapped. The additional spool travel needed to cross the overlap will cause a small step increase in pressure before reverse flow can occur. Alternatively, a Type B reducing valve will need a check valve in the mainspool to discharge leakage from inlet to outlet if the outlet port is blocked. The small pressure difference needed to open the check valve will appear as a small step-up in reduced pressure at the outlet port, figure 8.2.

It should be remembered that a combination of high flow and high differential pressure can inadvertently occur at the start of a cylinder stroke where there is no resistance to flow and cylinder pressure is very low. Under these circumstances a Type A reducing valve may close reducing the outlet pressure to near zero.

If flow demand at the outlet of any reducing valve exceeds flow available at the inlet to the reducer, the valve will attempt to increase reduced pressure by opening the mainstage. The result of this will be a loss of inlet pressure as the reducer will take all the available flow.

3.8 Linearity

Linearity of the demand signal to pressure characteristics of a pressure control valve without amplifier can vary between 3% and 10%, figure 8.3. A substantial proportion of the non-linearity will normally come directly from the magnetic characteristics of

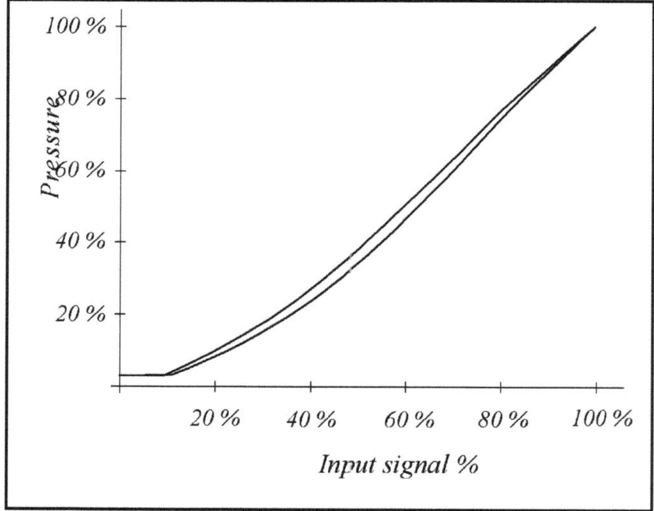

Figure 8.3: Typical Signal/Pressure Characteristics

the solenoid. Linearity of a relief valve should be measured with a fixed flow through the valve and consequently, zero signal cannot give zero pressure due to minimum head loss in the valve.

The addition of correction circuits to the amplifier can compensate for valve non-linearity over the majority of the flow and pressure range. Some manufacturers offer valve and amplifier packages which provide an overall linearity of about 1%.

Use of a stroke controlled solenoid will give a similar improvement in linearity but the low-pressure deadband may be larger with such valves.

3.9 Hysteresis and Dither

Hysteresis in a pressure control results from a combination of both mechanical friction and magnetic hysteresis of the solenoid or force motor.

The pressure of hysteresis is evident when the same current is approached from different directions and will typically result in a pressure difference of between 1% and 6% of the maximum pressure. If the direction of approach is consistent and the current profile is repeatable, then after a few cycles no magnetic hysteresis will be observed but pressure scatter is still possible due to mechanical effects.

The pressure of hysteresis is also shown when the same flow is approached from different directions. Hysteresis due to current and flow changes may be superimposed in use but each may be tested for and observed separately. If one figure is quoted it is normally hysteresis due to current.

The performance of all electrohydraulic pressure controls can be improved by the use of dither in the current signal. This is a very effective method of reducing both mechanical and magnetic hysteresis.

The dither amplitude and frequency must be chosen carefully to be sufficient to avoid excessive vibrations in the system while moving the valve enough to reduce hysteresis. This is often best achieved by testing on the application. The use of dither may increase the mean current at zero signal level and result in a slight increase in minimum pressure.

4 DYNAMIC PERFORMANCE CHARACTERISTICS

4.1 Frequency Response (Current to Pressure)

Under ideal test conditions with a very small volume of oil under compression, a single-stage valve may achieve a bandwidth as high as 80 Hz, while a two-stage valve may reach about 40 Hz. In practical circuits the effect of trapped volume will be to slow down the response because of the flow needed to compress and decompress the oil in the system. Typical installed frequency response will be between 5 and 20 Hz, except for very small or very large systems.

Some valves may be limited in frequency response by the amplifier or mechanical construction. In particular, single-stage valves with a position controlled solenoid may be limited in frequency response due to the time to move the solenoid.

4.2 Frequency Response (Flow to Pressure)

The current pressure frequency response is not always a good guide to a valve's ability to cope with oscillations in the flow. Response to flow oscillations will be a function of override and inherent damping within the valve. Response will still be a function of trapped oil volume and maximum bandwidths will be as stated above.

Valves with position controlled solenoids may give a greater bandwidth when subject to flow oscillations as response time of the solenoid is eliminated.

4.3 Step Response

A pressure control may be subject either to a step change in demand signal or step change in flow. Under either condition a major part of the total response time for a large step will be the time to compress or decompress the oil in the system. The valve will often fully open or fully close until the pressure approaches the valve's pressure setting. The response time of the valve on its own may be as short as 10 ms but is difficult to measure in isolation.

4.3.1 Step Response (Relief Valves)

If a relief valve is subject to a step change in flow which produces a small step change in pressure, i.e. less than 50 bar, there will normally be little time for the valve to respond and overshoot can be as high as 50%. As the magnitude of the pressure change increases, it is normal for the overshoot to reduce and above 250 bar there may typically be no overshoot.

Step changes in demand signal are less likely to produce overshoot and ramps in the amplifier can be used to reduce overshoots if necessary.

When a relief valve is subject to a negative demand step, the rate of pressure change is not dependent on the pump flow but the peal flow rate possible through the valve. This can be very high and to avoid tank line shocks, a ramp in the amplifier for negative steps is often desirable.

4.3.2 Step Response (Reducing Valves)

The step response of a reducing valve depends on trapped volume of fluid at both inlet and outlet ports and the nature of the applied step. Step changes to the flow at the inlet, outlet and in current will all cause different results, as will the steady state flow through the valve when the step is applied.

The most significant differences in performance occur when a reducing valve is subject to a step increase in flow through the valve. Fast valves may show little undershoot and recover in about 20 ms. Slower valves may allow reduced pressure to fall to near zero for a short period and take up to 80 ms to recover to operating pressure.

If the outlet flow is suddenly reduced to zero, a pressure overshoot will occur and fluid trapped in the outlet volume must be discharged to tank via the valve to recover to the correct pressure setting. If the valve has limited reverse flow capacity (see 8.1.2), the time to recover may be a few seconds, although this could be shortened by a temporary reduction in electrical signal to the valve.

5 RATED PRESSURE

The rated pressure is the maximum safe working pressure of the valve and is based on the strength of the pressure containing envelope.

The manufacturer will normally have completed fatigue tests to show that the valve is suitable for use in normal applications up to the rated pressure. In use all pressures applied to the valve, including overshoots, should be less than or equal to the rated pressure.

It is normal for the pilot drain or tank port of an electrohydraulic valve to be rated at a pressure lower than the rating of the inlet ports. This is seldom a problem as few applications require pressurised tank lines.

6 ELECTRICAL INTERFACE

To obtain the specified performance from any electrohydraulic valve it is important that the electrical drive conforms to the manufacturer's recommendations.

Electrohydraulic valves without feedback are usually controlled by current rather than voltage as this reduces the valve's sensitivity to temperature changes in the coil but the design of the current drive amplifier is critical to the speed of response and accuracy of the valve. For example, if insufficient voltage is applied the rate of increase of current will be limited and consequently, the speed of response will be less than expected, particularly if the valve has an electrical position or pressure feedback. It is usual then for the manufacturer to offer a drive amplifier to make correct use of the feedback and provide the often complex electrical functions necessary if the valve/amplifier pair is to give the performance given in the catalogue. Such amplifiers often provide additional functions such as input gain adjustment, ramps, offsets and detection of loss of feedback signal.

7 REFERENCES

ISO 10770-3 Test Method for Electrohydraulic Pressure Control Valves (under development)

ISO 6403 Hydraulic fluid power - Valves controlling flow and pressure - Test Methods

ISO 4411 Methods for determining pressure differential/flow characteristics

BFPA/P5 *'Guidelines to Contamination Control in Hydraulic Fluid Power Systems'*

THE APPLICATION OF ELECTROHYDRAULIC VALVES

1 INTRODUCTION

Over the past decade, the range of electrohydraulic proportional control valves has expanded considerably and now covers a wide spectrum in performance. In their simplest form they are basically flow or pressure control valves with the addition of a solenoid to provide remote operation and a performance comparable to manually-operated valves. At the other end of the spectrum they have a high dynamic performance which is suitable for the most demanding closed-loop applications. In between these extremes, the valves' performance and price can be chosen to meet the needs of a whole range of different applications.

Control has become a growth area in hydraulic applications, although its impact has varied between different industrial sectors. Some are more advanced than others which is more a reflection of less pressure on the latter sectors to improve, rather than of the inability of hydraulics to make the improvements.

In the following sections the applications have been sub-divided into different industrial sectors. Within each of these sectors the breadth of activity is outlined and then a few examples considered in more detail.

2 MANUFACTURING AND PROCESS CONTROL

Hydraulics are used extensively in the manufacture of materials and the subsequent processing of materials into consumer and industrial products. Plastic processing machines such as injection and blow moulding machines have traditionally been based on hydraulic systems. The use of electrohydraulic valves has reduced cycle times and improved quality. Reduced cycle times result from being able to open and close the heavy mould plates of the injection moulding machine faster, with less shock and associated risk of mould damage. Improved quality control comes from better control of the injection of the plastic into the mould. In particular electrohydraulic valves allow the velocity of injection to be profiled during the injection stroke to suit the shape of the mould. This can assist in eliminating voids and residual stresses in the moulded component.

Electrohydraulic valves allow the continuous adjustment of plastic film thickness produced by an extruder by setting the gap between the dies. Extruders are used to produce the small tube, often called a parison, that is blown into a plastic bottle. Adjustment of wall thickness of the parison over its own length is important for controlling wall thickness of the blown bottle and consequently is part of the quality and cost control process.

Electrohydraulic valves have made possible a new generation of pipe bending machines such as those used in the manufacture of automotive exhaust pipes. The addition of a feedback sensor to the actuator eliminates the need for mechanical end stops or travel limits required to set the angle of bend. With these end stops eliminated, production can change from one design of pipe to another almost instantaneously. The control of bend rate improves product quality and higher production rates are possible.

The ability to supply a high force from a small actuator has resulted in substantial use of hydraulics on machine tools. The stiffness and force of a hydraulic actuator allows the hydraulics to move the tool onto the work piece while it is being cut. Hydraulic power is ideally suited to all forms of clamps because of the simplicity of clamp design, the ability to limit clamping force by limiting pressure and the low or zero power usage during clamping.

Metal presses, particularly the larger ones, have traditionally used hydraulics. Improved control of force and velocity is being achieved by electrohydraulic valves as described below.

2.1 Force Control Presses

A press can be used for forming metals or plastics but can also be used for clamping laminates while they cure. In both cases control of the force output from the actuator is desirable and can significantly influence the quality of the final product. In practice it is often difficult to measure force with a load cell due to mechanical construction or the magnitude of the force generated. Fortunately, in cylinder applications the area subject to fluid pressure is constant and consequently the control of pressure in the actuator is often a satisfactory method of controlling the force.

The traditional method of controlling pressure has been by the use of a hydro-mechanical relief or reducing valve. These valves can be replaced by electrohydraulic versions but system performance may not be improved since pressure can still change as flow through the valve changes. To improve accuracy or repeatability of the pressure setting a pressure transducer can be used. The signal to the electrohydraulic relief or reducing valve can be adjusted to obtain the desired pressure accuracy. A suitable amplifier and electrical closed-loop is required for such a configuration.

During the cure cycle of a laminate press very low forces and hence low cylinder

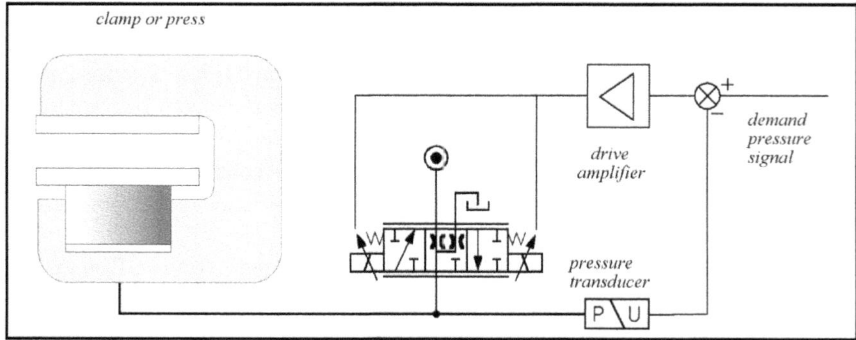

Figure 9.1: Closed-loop Pressure Control Circuit

pressures may be required. Sometimes relief and reducing valves are not able to control down to low enough pressures. However, by using the arrangement shown in figure 9.1 a direct acting electrohydraulic spool valve is able to provide pressure down to 0.1 bar provided the pump flow is small enough relative to the valve flow rating. This configuration requires a direct acting valve in which the solenoid is able to push the spool to its fully open position. In addition, to obtain accurate pressure control the spool valve should have high dynamic performance and be zero or slightly underlapped. If these requirements are met, excellent accuracy and repeatability can be obtained over a wide range of pressures.

2.2 Speed and Force Control in Plastic Processing

In an injection moulding process the plastic must be injected into the mould in a short time. To do this, the inject screw is moved rapidly forwards in its barrel without rotation. This linear, inject movement is typically achieved using one or two hydraulic cylinders as shown in figure 9.2. Electrohydraulics can improve the quality of injection moulded components by controlling two aspects of the injection process.

Firstly, during the injection of plastic it is desirable to control the speed of injection into the mould typically reducing speed as the plastic flows through small sections of the mould. Secondly, to stop shrinkage, some force must be used to pack the plastic into the mould but this force must be gradually reduced as the plastic cools.

Speed of injection can be controlled by a closed-loop system using inject cylinder position as a feedback signal, as shown in figure 9.2. Alternatively, as the flow into the

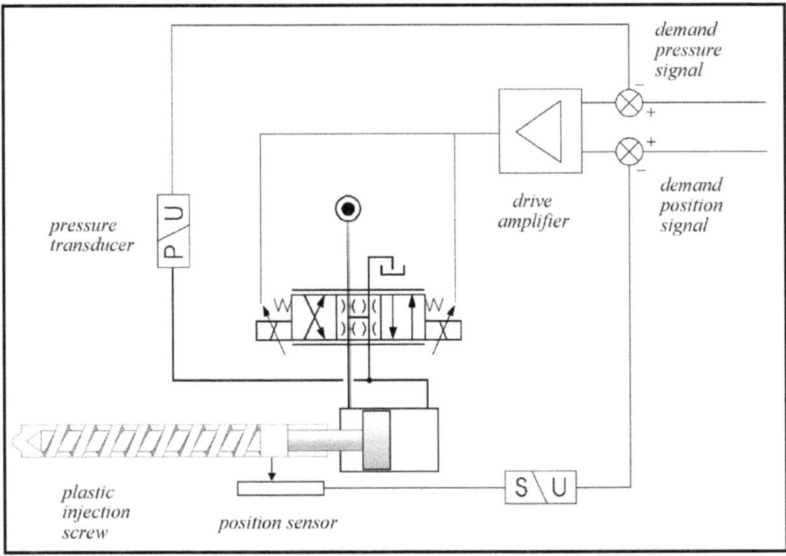

Figure 9.2: Closed-loop Pressure and Stroke Control Circuit

inject cylinders is proportional to the velocity the flow may be controlled instead. Provided the pressure drop across the electrohydraulic valve is substantially constant, spool position control can be used instead of a flow signal. The system accuracy requirements will dictate which method of controlling velocity is used.

To ensure the mould is completely full, the inject cycle must attempt to inject more than enough plastic into the mould. At the moment the mould is completely full, the speed control system naturally drives the pressure up to try and maintain the programmed speed. To obtain a good injection moulding it is important to limit pressure overshoot under these conditions.

The system in figure 9.2 uses the electrohydraulic valve to control both velocity and pressure. This arrangement can provide advantages in installation space, cost and control accuracy compared to two independent valves. To implement such a system, pressure must be measured with a transducer and the amplifier controlling the valve must switch to pressure control mode at the appropriate time.

Valves used to control both pressure (P) and flow (Q) are sometimes referred to as PQ valves. Such valves may incorporate an electrical pressure sensor and may use a spool position sensor to give an approximation to the actuator velocity. The PQ valve hardware will be very close to a conventional electrohydraulic valve, except for a special spool with a zero lapped pressure control position.

2.3 Rug Manufacturing

As a final example of electrohydraulics in manufacturing the improvements to a rug making machine are described. This application demands high dynamic performance and positional accuracy. The rug making industry uses a "Tufting" machine in the manufacturing of "shag" type rugs. A pattern can be generated on the rug face by positioning needles which insert yarn into a backer material. To improve life and speed of manufacture an ultra-life electrohydraulic actuator was developed as a "Crossover Needle Bar Positioner".

The requirements for the system were to position the needle bar to an accuracy of 0.15 mm following a 9,5 mm step within a time period of 0.0565 s, under a load of 533 N with an overall stroke of 50 mm. The operational rate was to be 7.5 stitches/s.

The transient response requirements together with the need to reduce errors due to valve hysteresis and null shifts led to the need to use a high loop gain. These conditions could only be satisfied using an actuator with a load resonant frequency in excess of 130 Hz together with a high response electrohydraulic valve. The electrohydraulic

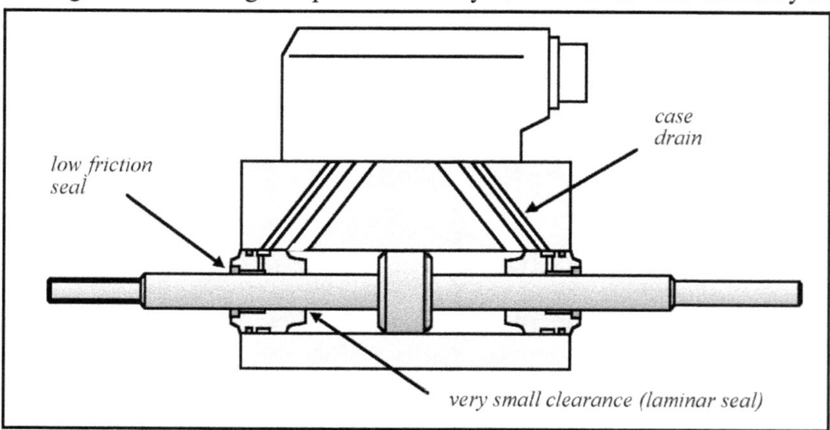

Figure 9.3: Low Friction Actuator

valve is mounted directly on the actuator which minimises the oil volumes in the interconnecting lines and hence minimises their effect on the load resonance, see figure 9.3. An LVDT transducer, with an overall stroke of 100 mm, was used to measure the actuator position.

The long life of the actuator was obtained with the use of non-contacting, laminar, high pressure rod seals which have a tightly controlled clearance to maintain an oil film between the seal surface. The piston head and rod are a one piece design. The piston and rod are made from stainless steel and the actuator body from aluminium. The rod end seals are match fitted with typical diametral clearances of 0.005 mm. As any leakage outside the actuator is not tolerable, low-pressure lip seals are added to the piston rod on the low-pressure end of the rod seal and the leakage is returned to the reservoir via a separate case drain. The case drain leakage for a new actuator is typically about 0.0025 L/min.

3 MECHANICAL HANDLING

For many people today mechanical handling is synonymous with robots. Robots, with their flexibility of movement, are often seen as the future for all mechanical handling but many conventional systems such as the ubiquitous forklift truck will provide mechanical handling functions for many years to come. Electrohydraulic drives have increased the capacity and reliability of conveyors. Automatic control of the belt drive reduces the risk of damage to the conveyor by, for example, ensuring all speed changes are very gradual.

Automated warehouses often use unmanned trucks, similar to forklift trucks, to store and retrieve items as required. Hydraulic systems have proved to be very suitable for these trucks where the stiffness of the hydraulic control system gives the reproducible positioning required for loading and unloading even when handling a wide range of loads.

In manufacturing, special purpose machines, often based on hydraulics or pneumatics, still predominate for high speed repetitive tasks such as electronic circuit board assembly. Where space is at a premium the actuators can be built into, or be part of the machine. Improved magnetic materials are making electric drives more competitive in small systems but for larger machines, the greater force capability and improved output stiffness gives hydraulics the advantage.

3.1 Lift/Lower Controls for Lift Trucks

A critical function in a lift truck used for automated warehousing is the remote control of the lift cylinder. The circuit in figure 9.4 shows how the lift cylinder can be controlled by an electrohydraulic valve while meeting the following operational requirements:
- to provide maximum efficiency, requiring the pump to be on only while raising the load and to use the stored energy in the cylinder for proportionally lowering the load.
- to provide positive load holding, manual lowering and other safety features needed in this application.

Lowering the load under electrical control requires the solenoid controlled check valve to be energised and a command signal to the electrohydraulic valve to move the

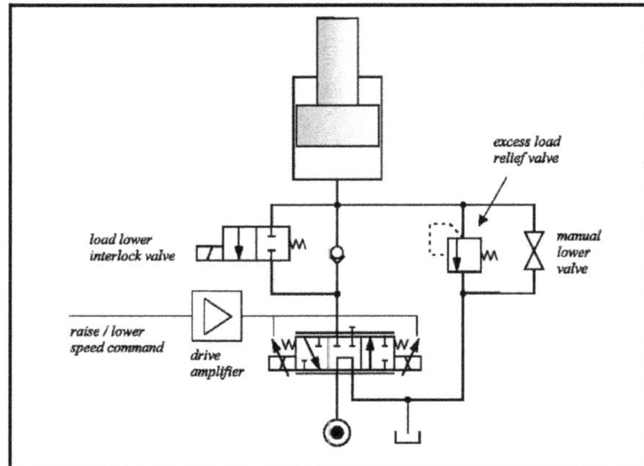

Figure 9.4: Lift Truck Circuit

spool to the right, as shown in figure 9.4. The pump remains off during the lower mode. The velocity during the lowering mode is a function of both command signal and load pressure. If required, the variation in speed due to load pressure can be reduced by flow controls. Such flow control valves may be located in the base of the cylinder to prevent excessive velocities if the hose to the cylinder should break.

Lift mode operation starts when the electric motor driving the pump is turned on. At this point, the pump is generating full flow and the spool in the electrohydraulic valve is still in its centre position. The design of the valve spool allows pump flow to be discharged to tank at a nominal pressure drop of 20 bar.

An increasing command signal is applied to the electrohydraulic valve moving the spool gradually to the left. As the spool moves it restricts the pump flow path to tank, thus raising pump pressure. At the same time it opens a metering slot to connect pump pressure to the lift cylinder. When the pump pressure reaches the lift cylinder pressure, typically 35 to 175 bar, the load begins to rise. Further movement of the spool opens the metering slot allowing a greater percentage of the flow to reach the cylinder. If the spool is moved fully to the right the cylinder receives 100% of pump flow and pressure loss through the valve is minimised.

Thus, the most inefficient portions of any life cycle are during the acceleration and deceleration portions of the lift. One of the obvious advantages of electrical control is that the command signal can be shaped to minimise acceleration and deceleration times within the limits of valve response time, operator comfort or other smoothness requirements.

3.2 Glass Transporter

Extra large forklift trucks are used to transport glass during its manufacture. Plate glass is made in sheets 2.5 m wide and 6 m long. They are stacked on carrying frames which have a 30 tonne payload. The glass carrying frames need to be moved within the production plant to where they are needed and then set down in an accurately defined position so that sheets can be picked up by a robot feeding the glass to the next manufacturing stage. All the hydraulics on these special transporters are controlled by a programmable controller allowing one man to operate the truck.

Hydraulic power provides the muscle to lift the glass and move the transporter. The carrying frame containing the sheets is lifted evenly by four cylinders with built-in position transducers. Electrohydraulic valves control the cylinders so that the synchronisation error does not exceed 3 mm over a stroke of 750 mm. A hydrostatic transmission is used to drive the vehicle via wheel motors located in the front hubs and power braking is implemented using a separate circuit.

4 AUTOMOTIVE AND RAIL TRANSPORT

Special purpose hydraulic systems have been used in road vehicles for many years. With the advent of affordable control valves both the number and sophistication of the systems are increasing. Power steering is a case in point. The original simple systems are being improved and in some cases, extended to four-wheel steering by the use of proportional valves. Powered braking systems are replacing vacuum operated ones. Power hydraulic braking using electrohydraulic control has advantages over vacuum operation particularly for anti-lock systems.

Active suspensions were shown to be very advantageous in Formula 1 car racing where road holding is paramount. For passenger vehicles, ride is equally important and some luxury cars now have variable ride control as standard. Interest in this technology is high and further developments can be expected. These systems exploit the high power-to-weight ratio of electrohydraulics which cannot be matched by any other technology.

Active ride control systems are not limited to road vehicles. Systems for passenger trains are expected to enter service soon. For any ride control system best results are always obtained using an inertial reference and a fully active hydraulic support system. Such systems are typically expensive and can absorb a lot of power. Very similar results, in terms of passenger comfort, can be obtained by a double rodded cylinder with an electrohydraulic throttle valve connected between the two ends of the cylinder. Signals to the electrohydraulic valve are provided by an electronic control box that has multiple inputs such as bogie acceleration.

Hydraulics have been used extensively in the support of rail services. In the UK the majority of points are now controlled remotely using small hydraulic power packs at each point to provide actuation. Maintenance trains operate with hydrostatic transmissions and use electrohydraulics for tamping sleepers into place and rail handling.

4.1 Engine Cooling Fan Drive

A simple application of hydraulics is an engine cooling control. Internal combustion engines will always require some form of cooling fan drive system. As space limitations are ever more a feature of vehicle design, alternative fan and radiator positions have to be considered. The combination of a hydraulic motor to drive the fan and an engine driven pump provides excellent flexibility in radiator location and allows the use of larger, power hungry, fans. This versatility has led to the system being extensively exploited by manufacturers of trucks, mid and rear engine buses, construction vehicles and rough terrain military vehicles.

This hydrostatic system includes an electrohydraulic valve that allows the speed of the fan to be selected, largely independent of engine speed. This is a significant advantage compared to mechanical systems in which fan speed adjustment is difficult. Adjustment of the fan speed provides a method of controlling the engine temperature. Temperature sensing can be by one (or more) thermistors with the control amplifier driving both the electrohydraulic valve and radiator shutters. Direct control from engine management systems may be possible at a later date.

The heart of the fan drive system is a special electrohydraulic valve mounted on the motor. This valve diverts flow around the motor and back to tank depending on the demand for cooling. In operation the pilot of this two-stage valve applies pressure to one end of the main spool while the other end is connected to motor inlet pressure. Excess motor inlet pressure causes the main stage spool to dump a higher proportion of flow to tank. The motor's pressure/flow relationship is dependent on the size of the fan being driven. Hence the spool is designed to meet the particular application requirements and parameters such as motor size, fan type and maximum fan speed.

5 MARINE AND OFFSHORE

In marine and offshore applications reliability is at an absolute premium, not only due to the difficulty of repair and maintenance but also because the cost of system failure can be substantial. Consequently, a conservative approach is adopted to the design of systems with the choice of components favouring those with a simplicity of maintenance and a proven track record for reliability.

Over the past decade there have been significant increases in the use of electrohydraulic valves. The key advantage of electrohydraulic valves is that they can be mounted on or close to the actuator, so eliminating the long hydraulic lines associated with manually controlled valves. Short lines to the actuator are advantageous because at sub-zero temperatures the oil in long deck-mounted lines can become very thick and consequently actuator response can be very slow.

Hoists and winch speed control systems benefit from electrohydraulic control as the characteristics can be electrically modified to improve ease of use. Winch systems operating with closed-loop control can maintain a fixed clearance between the load to the deck simply by reducing the signal specifying the clearance. Bow thrusters and variable pitch propellers also benefit from electrical adjustment of the response characteristics to make handling of the ship more predictable. Bucket dredgers also use electrohydraulic control valves in their operation.

Engine mounted hydraulic pumps can easily provide the very high power often needed in marine applications. Consequently other technologies can rarely compete for tasks like the control of ship stabiliser fins. The control of these fins requires significant forces and good position control, under conditions of rapidly changing disturbance forces.

5.1 Offshore Escape Gangway

Iolair is a dynamically positioned semi-submersible emergency support vessel stationed in the BP Forties field in the North Sea. Normally it provides support for diving operations and other maintenance tasks but in an emergency it is able to provide powerful water jets for fire fighting and evacuate personnel from a platform via the hydraulically positioned gangway.

The gangway that allows personnel to walk from the platform to Iolair is designed to operate in wind conditions up to Force 9 and the associated heavy swell. To perform this task all movements of the gangway, including elevation, rotation and telescopic extension, are hydraulically controlled.

The hydraulic system gives the extension movement and controls the gangway length. It comprises of two winches, each operated by a bent axis motor and an electrohydraulic valve. Relief valves are incorporated between the motor service lines to limit the pressure drop across the motors for rotation in both directions. The electrohydraulic valves are rated at 230 L/min.

The whole system can be locked by the de-energising of two intrinsically safe solenoid valves. They system will only be unlocked when these valves are energised allowing a pilot pressure to operate the check valves. This pilot pressure is also cross connected between the two winches so that, in certain cases, the extension may be driven by a single winch. A manually-operated ball valve allows the service lines between the motors to be connected if required. This valve would be kept locked in normal use.

5.2 Stabilisation of a Ship-Mounted Derrick

The bottom of the Pacific Ocean, some three or four miles deep, is covered with mineral rich nodules which are being pumped to the surface with a pipe and dredge-head system. The pipe is deployed by a ship-mounted derrick, which is gimballed on a large elastomeric bearing and is tilted relative to the ship by four high performance actuators controlled by electrohydraulic valves.

Due to the length of the pipe and its weight (up to 550 tonnes), the pipe stresses are considerable and even a small amount of local bending in the pipe-string would cause it to break. As the ship can pitch and roll by as much as 12 degrees in a heavy sea, the mast was mounted on top of an elastomeric bearing and held vertical by four hydraulic cylinders, one at each corner of the mast base as shown in figure 9.5.

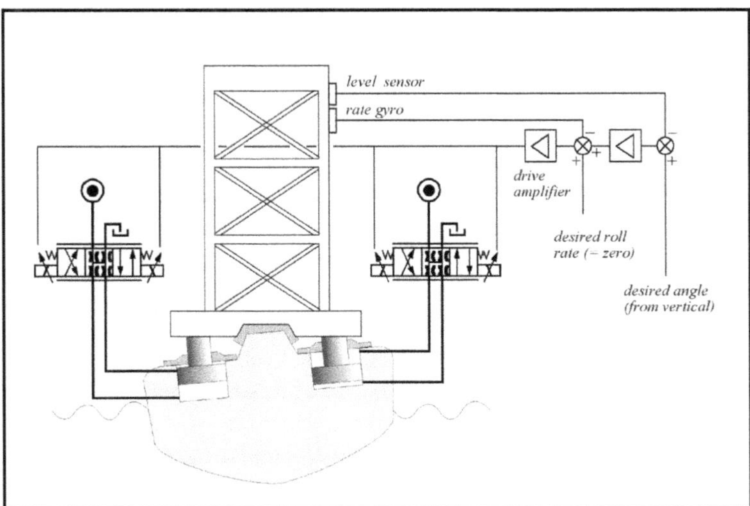

Figure 9.5: Angle and Roll Rate Control Circuit

There were two primary design requirements for the control system:

- The mast, 21 m high and weighing 200,000 kg, must be held vertical within 1 degree under normal operating conditions.

- The system must continue to operate normally after any failure long enough for the crew to terminate pipe-handling operations.

Normal operating conditions were defined as a pitch and roll frequency of 0.5 rad/s and horizontal accelerations of 0.12 g's. The centre of gravity of the mast was 12 metres above the centre of rotation and its moment of inertia about the centre of rotation was 29 x 106 kgm^2.

The system was successfully implemented by using two axes of control, each axis being driven by a diagonally opposite pair of hydraulic cylinders 4 metres apart, each cylinder having a piston area of 460 cm^2, primary electrohydraulic valves and a backup valve with its own drive electronics. Each electrohydraulic valve has a flow rating of 780 L/min at 70 bar while total flow was 1800 L/min at 210 bar.

To obtain the specified performance a rate gyro and a pendulum sensor were required for each axis. The highly sensitive rate gyro loop was used to hold the mast in a fixed angular position while the average output from the pendulum sensor was used to correct drift in the rate gyro loop and hold the mast in a vertical position.

6 MOBILE EQUIPMENT

Mobile equipment, which includes a wide variety of on and off road machinery, is one of the largest sectors of use for hydraulics. In many of the machines such as gritting vehicles and forestry harvesters the electrohydraulic control valves replace manual direct operated valves giving advantages in location and sometimes improved control characteristics. Other machines exploit the closed-loop capability of these valves, examples include road graders and grain harvesters.

Electrohydraulic valves provide advantages to the operators of mobile machinery by reducing the noise generated in the cab. The use of manually-operated valves demands that hydraulic piping is routed to the cab and back to the actuator. The use of electrohydraulic valves eliminates the multiplicity of pipes and the attendant noise. For the machine builder the reduction in piping complexity can easily offset the increase in valve cost. In addition, ergonomically designed controls can be positioned to relate better to the motion under control, e.g. left/right swing.

A factor common to many mobile machines is the need to operate drive transmissions in a flexible manner to meet the duty requirements, particularly for tracked vehicles. Hydraulic systems inherently have advantages over mechanical transmissions and proportional control can further improve flexibility.

Hydraulic excavators, particularly those based on wheel tractors, are the most commonly used pieces of mobile plant. A factor dominating their design is that they are frequently hired by contractors. Consequently the need to be very simple, extremely robust and able to tolerate a considerable degree of misuse. The majority of machines are based on direct manually-operated valves. However, this situation is changing. Systems using proportional valves are being developed for the automatic control of the excavator digging cycle. Here the operation of three separate hydraulic systems need to be synchronised. Automatic control improves the digging efficiency, particularly compared to that achieved by inexperienced operators.

6.1. Access Platforms

Access platforms exploit the power to weigh advantages of hydraulics. Fluid power is easily transmitted to each joint where an electrohydraulic valve and actuator can adjust the angle of the joint. Electrohydraulic valves give flexibility of control as the platform position can be adjusted from controls on either the platform itself or on the vehicle. One of the latest developments in access platforms is claimed to be the highest self-propelled lift with a range of platform height from 45.7 m above ground level to a depth of 9 m below. The lift movement is obtained through a three section telescoping boom with separate swing and drive systems. PQ valves are used to give the degree of control required, particularly in achieving both ramp and creep control.

6.2 Electronic Load Control of Excavators

Excavators rely entirely on hydraulic power to achieve their function. In most cases a diesel engine drives a hydraulic pump with a hydro-mechanical load control, which provides services to the excavator arms, track drives and swivel. One of the main problems is to run the diesel engine at maximum efficiency while avoiding the risk of stalling. For any position of the accelerator pedal, a typical diesel engine produces its optimum power at an engine speed slightly below the no-load speed. Ideally the hydraulic load control system should continuously adjust the pump power requirements so that the engine runs at its optimum speed rather than simply limiting the power required for the pumps. In this way the control system can automatically compensate as the engine runs in or wears out. The use of electrohydraulic valves allows such a system to be implemented as shown in figure 9.6a.

During commissioning the electronic controller senses the setting of the accelerator pedal and the no-load engine speed. From this data the controller calculates the relationship between the two and establishes the optimum speed, often 85% to 90% on no-load speed, see figure 9.6b. In operation the load produced by a hydraulic pump is a combination of pressure and swash angle. An increase in pump load causes a reduc-

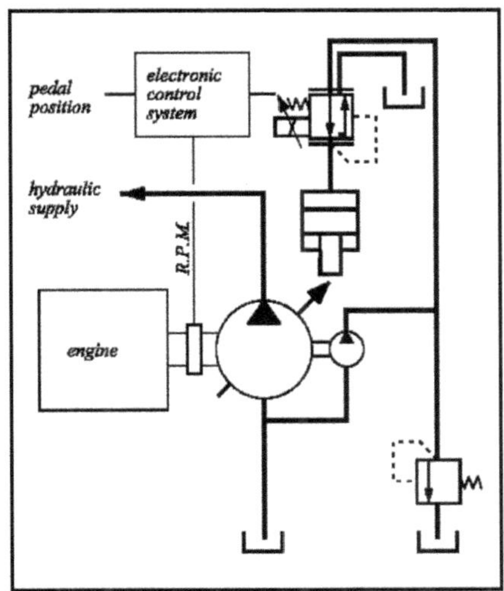

Figure 9.6a: Pump Swash Control

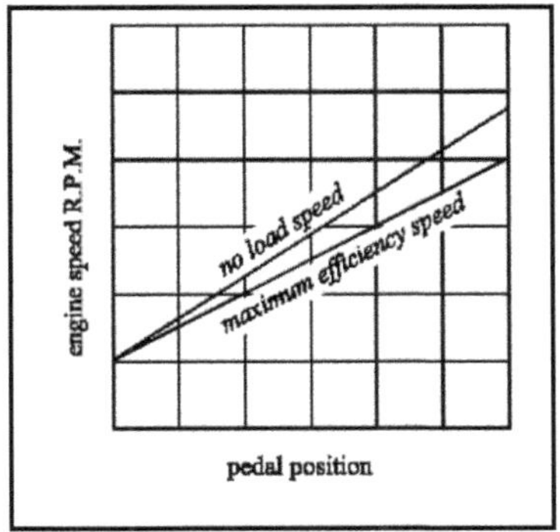

Figure 9.6b: Ideal Engine Speed

tion in engine speed. If engine speed falls below its optimum the electrohydraulic control system reduces the pump swash and thus power requirement until the engine speed rises to the optimum as defined by the controller. Alternatively, if the engine speed is above the optimum, the pump swash can be increased.

A further advantage of this more advanced control system is the ability to set up economy modes of operation where the preset speed drop is set to only a few percent which limits the power output of the pump set. The machine operator is given the option of full power or economy operation. Other options can be added including an engine over-temperature power reduction and a pump de-swash function if the operator does not use the hydraulic services after a preset period of time, typically 10 to 15 seconds.

7 MINING AND TUNNELLING

Machinery that is used for open cast mining is virtually the same as that used for earth moving. Tunnelling machinery is more specialised and the design depends on the material to be removed. The tunnel under the English Channel has been cut through relatively soft material which has allowed the use of a large single rotary cutting head. Machinery behind the cutting head is required to continuously remove spoil and automatically position roof support segments. Electrohydraulic valves assist in the accurate placement of these roof support segments.

Road headers use one or more smaller drills for tunnelling in hard rock. These have rotary cutters on the end of a jib and multiple holes are required to describe the tunnel profile. Accurate positioning of the holes reduces the amount of concrete needed to make a smooth wall inside the tunnel. Electrohydraulic valves are used in closed-loop systems to improve position control of the jib and the accuracy of the tunnel profile. The saving in concrete costs in a few miles of tunnelling can easily pay for the increased cost of the machine.

Drilling for blasting and roof bolting is a very common operation. Electrohydraulic valves are used to orientate the jib to the angle required and improve the positional accuracy of the hole.

Hydraulics are used for many systems concerned with the operations of mines. Hydraulic conveyors, some of which include belt weighing systems with electrohydraulic valves, are very widely used. Hydraulically operated bunkering systems make use of the belt weighing by removing excess coal when this would exceed shaft capacity and loading it up later when the conveyor delivery reduces.

7.1 Mining Haulage Control System

This is a nominal 300 kW system controlling an endless rope haulage. It is used for moving men or materials over approximately 2,400 metres and provides a rope speed of about 18 km/hr. It is capable of a maximum pull of 9 tonne and can operate on a 1:10 gradient.

The system employs two variable delivery pumps which are used in a closed circuit hydraulic system to drive a single hydraulic motor which is mounted inside the surge-wheel. The fluid used is a 60/40 water-in-oil emulsion and the maximum system pressure is 170 bar. The speed of the fixed displacement motor is controlled by pump delivery. The hydraulic pump delivery is adjusted by an electrohydraulic valve controlling the pump swash angle. The swash control system requires a separate hydraulic supply and an auxiliary pump is provided for this purpose. The pump swash control mechanism is spring centred and requires a pressure between 10 and 45 bar to vary the output from the mid position "no flow" state to maximum flow output. The swash can be moved to either side of centre, reversing the pump output ports and thus reversing the haulage.

The driver, travelling on the train driven by the haulage, operates the system remotely via a radio-controlled, intrinsically safe, electrohydraulic valve. Pressure transducers adjacent to the actuator provide the feedback signals required to form a closed-loop system that ensures the correct pressure is maintained. The haulage has a fully automated brake and speed overriding control, essential in this particular environment, so that the driver's signal can be overridden in cases of emergency or if excessive speed has been demanded.

8 CIVIL ENGINEERING

There are two aspects to civil engineering equipment, the construction phase which normally uses mobile and other special purpose plant and the subsequent operation phase. The use of hydraulics in the latter is mainly limited to the movement of large structures - flood gates such as the Thames Barrier, lock gates and road bridges. An example of the latter is a bridge in Dublin.

8.1 Lift Bridge

The road bridge over the River Liffey in Dublin is a single span which lifts to allow shipping to pass. It is a single sided rolling bascule bridge carried on two quadrants, 2 m radius, with parallel running gear racks, see figure 9.7. It is some 40 m long, weighs 450 tonnes and has a single carriageway and footpath in each direction.

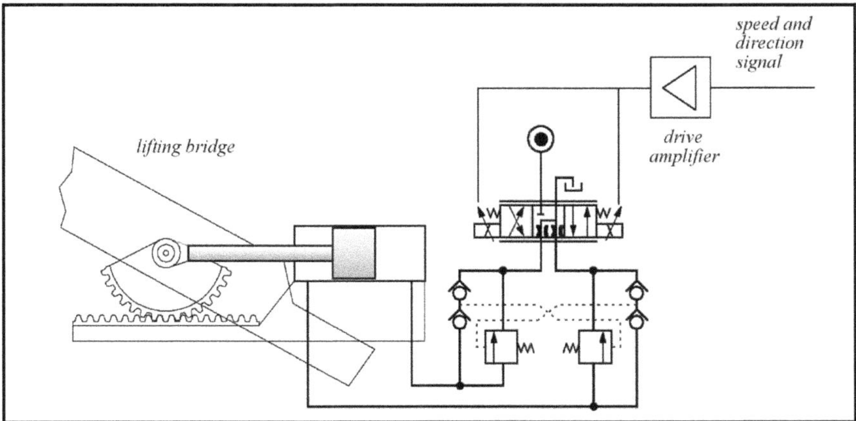

Figure 9.7: Bridge Lift Mechanism and Control Circuit

In addition to the requirements for locking the bridge in both the upper and lower positions and requirements for emergency operation, the specification included bridge operation at wind speeds of up to 100 km/hr and a normal lifting time from down to fully up and locked within 55 seconds. Allowing for acceleration and deceleration times this gives a tip speed of about 1.1 m/s. Cylinders of 320 mm bore and 140 mm rod diameters with a 2.9 m stroke were selected which operate directly on the centre point of the rolling quadrant.

Two pumps driven by 75 kW motors were used for the main power, whilst bridge speed and directional control was via a proportional valve. The use of two separately driven main pumps plus two separately driven pilot pumps allows for redundancy under emergency conditions. The pressure setting of the control is normally 110 bar with an override control on the operator panel allowing the pressure to be raised to 150 bar for high wind operation. The hydraulic lock cylinders at the top and bottom of the bridge stroke are normal solenoid-operated directional valves.

9 METAL INDUSTRY

The environment of mills and foundries, often dirty and usually hot with attendant risk of fire, is an arduous one. Hydraulics have a track record of operating satisfactorily here. With increased use of continuous processes, particularly continuous casting, there is a greater emphasis on automatic systems to achieve both continuous performance and improved quality control. Hydraulic systems play an important part in all aspects of production. In the initial phase of manufacture, hydraulic systems control the walking beam furnaces and continuous strand casting machines.

Once the sheet or strip has been formed it may be rolled to reduce its thickness and improve its mechanical properties. Automatic thickness control is often used, taking the feedback signal from the measured sheet thickness in order to control the mill gap.

A variety of different systems are involved in handling the strip once it has been finished depending on the size and flexibility. Thin sheet is often coiled for easy transportation. For coiling the leading edge of the strip needs to be picked up by one system and fed into the coiling machine. When one coil is complete the sheet is cut by flying shears which travel at the same speed as the strip to produce a right angle cut and a new coil started. These operations are all performed high speed and need high reliability so that mill operation does not need to be halted.

9.1 Strip Edge Guidance

An edge guidance system using electrohydraulic valves can easily be built provided suitable sensors are available for sensing the edge of the strip or sheet being processed. For example, a non-contacting sensor that is suitable for hot or cold strip consists simply of a light intensity measuring cell. When positioned at the edge of the strip the cell provides an analogue signal proportional to the percentage of the light passing the edge of the sheet. A pair of such sensors can detect sheet position by taking the difference between the two sensors.

The difference signal, if desired, combined with an offset signal can provide the input necessary for the electrohydraulic valve responsible for adjusting strip position, see figure 9.8. A suitable difference amplifier is often available as an option on electrohydraulic valve amplifiers.

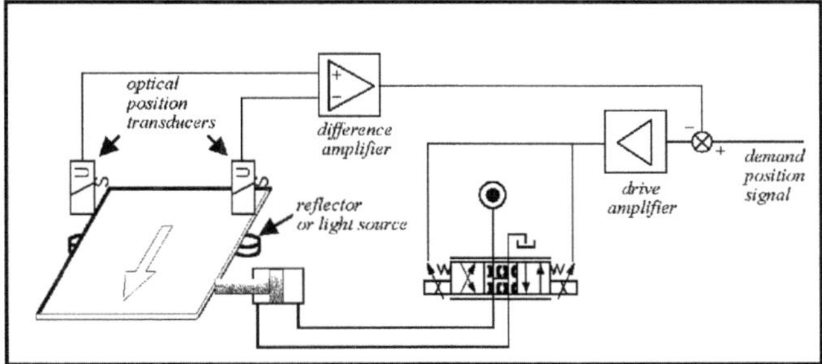

Figure 9.8: Edge Guidance Control Circuit

Stability of an edge guidance system for a continuously moving strip can be difficult to achieve if the sensors are not close to the positioning mechanism. The speed of the trip dictates the time delay between the movement of the positioning mechanism and the later movement of the strip at the position of the sensors.

9.2 Roller Position Control System

Obtaining accurate position control of a hydraulic cylinder is difficult if the load is highly variable. Rapid changes in load cause the oil to compress or expand before the closed-loop system has time to react. When rolling sheet metal, the load on the rollers can change rapidly and substantially due to variable initial sheet thickness. To obtain accurate thickness of the sheet being manufactured the position control system for the rollers must minimise any change in position.

Applying a constant back pressure to one side of the roller position control cylinder increases the stiffness of the system and so reduces position changes. The back pressure is easily generated by using a pressure control valve, see figure 9.9. Using the arrangement shown, the pressure control must be a type that allows flow into and out of the cylinders at sufficient speed for the maximum roller position adjustment speed required. However, the speed requirements for the roller position adjustment are typically very low. The use of an electrohydraulic pressure control valve allows the electronic system controller to adjust the pressure to suit the loads on the roller.

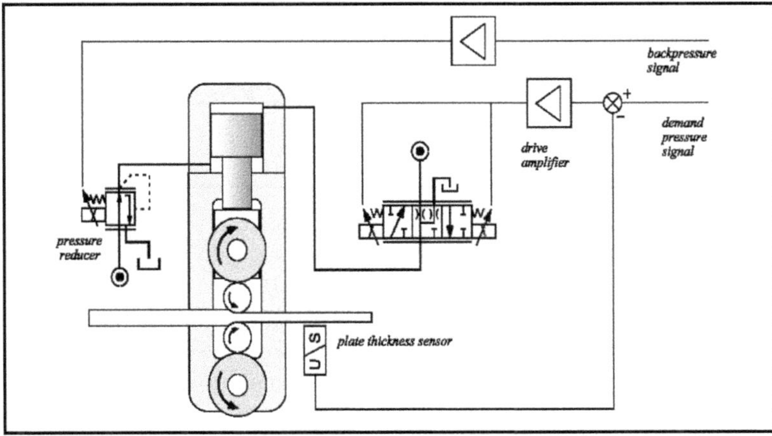

Figure 9.9: Roller Force Control

9.3 Continuous Strand Casting Plant

In these plants, a continuous strand of metal is pulled out of a mould. A hydraulic mechanism is used to do this which operates in a "hand-over-hand" manner. It not only needs to pull the strand outwards but also periodically produce a momentary reversal. This reversal has a major effect on the quality of the surface of the casting and it needs to be carried out in a very controlled manner.

The hydraulic mechanism uses two clamps and their sequence of operations is shown in figure 9.10a. Whilst clamp 1 is clamped on to the strand and is being pulled by withdrawal cylinder 1, clamp 2, which is not clamped to the strand, is being reversed. When withdrawal cylinder 2 nears the end of its stroke, its velocity is reversed until it is synchronised with that of clamp 1. Clamp 2 is then clamped to the strand so that its withdrawal cylinder is also pulling the strand. Clamp 1 is then released and returned to its original position.

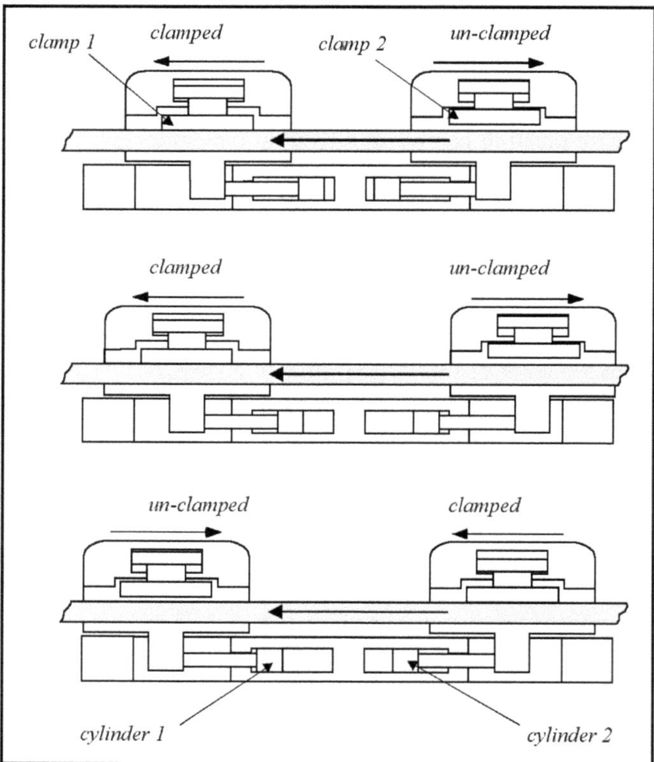

Figure 9.10a

This cycle of operation is repeated up to 4 cycles/s, with withdrawal cylinder velocities up to 7.5 m/min, see circuit diagram figure 9.10b.

Very good repeatability and accurate synchronisation are required to give a good quality product and this requires a high performance from the closed-loop system controlling the cylinders. Electrohydraulic valves are mounted directly on the cylinders to obtain the necessary dynamic response. The cylinder rods are supported in hydrostatic bearings, which also act as seals, reducing seal friction as far as possible. The pressure differential across the cylinder is measured, for calculation of the withdrawal force and this is also used in the control of the process.

The clamping cylinders incorporate inductive position sensors and are also controlled by electrohydraulic valves. The overall hydraulic power for these two systems, together with that for operating the guide rolls, is 300 kW.

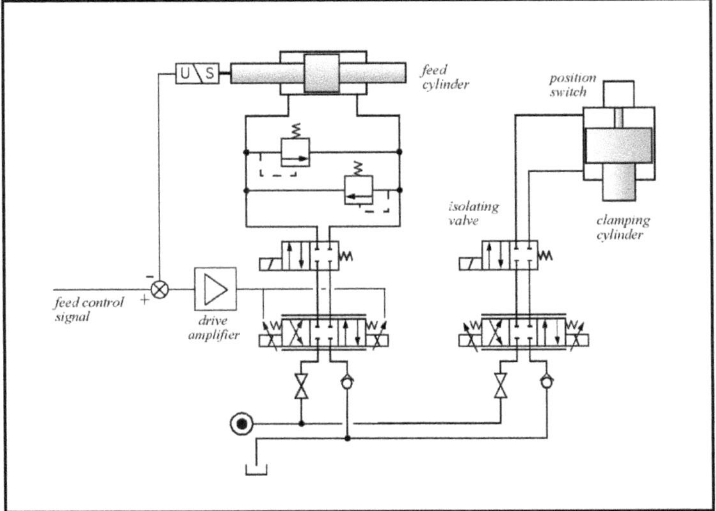

Figure 9.10b

10 THE TEST EQUIPMENT

A major consideration in the design of test equipment is the control of test parameters within close limits. A consequence of this is that a large majority of test rigs are based on closed-loop systems.

Health and safety considerations have increased the need for rigorous testing of a wide range of products and components. Test rigs using proportional or servo-valves with a good dynamic response can accurately reproduce special motion profiles tailored to meet individual test requirements. Such test rigs are significantly more useful than one driven by a mechanical cam or crank system. Vibration, fatigue and impact testing of mechanical structures will often exploit the capability of hydraulics to supply high force and instantaneous power. This latter capability being derived from the use of accumulators to store fluid under pressure and the use of high response electrohydraulic valves to supply the fluid to the test actuator. A good example of the use of this capability is the impact testing of cars by moving the hydraulically driven wall into the car rather than the car into the wall.

In some instances where large structures are concerned, models of the structure must be tested early in the design process. One example is the testing of models to estimate the wave loading on offshore structures. Test waves are generated in a plume or an open basin by large hydraulically driven paddles. These typically consist of a simple flap hinged to the bottom and moved backwards and forwards in the water. Although this is a simple arrangement, it is possible to reproduce not only peak wave heights but also a distribution of wave heights representing different sea conditions. Another method of testing structural models and their foundations is to load the model using a large centrifuge as described below.

A related area and perhaps one of the most demanding applications for any control system is a flight simulator. This involves the movement of a large mass, the simulated aircraft cabin, in three angular and three linear degrees of freedom. The peak velocities and accelerations are relatively large yet the motion at small velocities must be very smooth as the human operator can readily sense and be confused by any jerkiness in the motion which could be a consequence of the test gear itself rather than the simulated motion of the aircraft. The performance is achieved by the careful selection of valves and the design of the system to ensure that the interaction of the valve with the system does not produce these spurious effects.

In some industries the emphasis on quality is leading to comprehensive type testing before use and the continuous monitoring of component performance. In some applications the equipment must be very robust and be capable of use by relatively unskilled operators. The hydraulic damper test rig described below typifies the requirements and possibilities.

10.1 Hydraulic Damper Test Equipment

Hydraulic dampers are used extensively in automotive suspension systems and require rigorous type testing before use. Testing is undertaken at several preset velocities in each direction of travel. The force of up to 20 kN is required to reach the highest test velocity of 400 mm/s. Accuracy requirements demand the velocity is held within a 5% tolerance independent of load and a constant test velocity reached within the first 5 mm of travel.

A high performance closed-loop system, see figure 9.11, is needed to meet these requirements. Low friction seals are used in the actuator which has a built-in LVDT

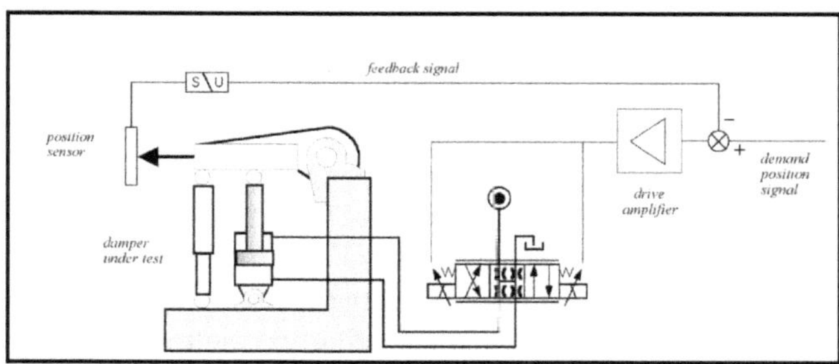

Figure 9.11: Automotive Suspension Test Machine

position transducer. The electrohydraulic valve, mounted directly on the actuator, has a frequency response of better than 100 Hz to achieve the system accuracy requirements.

Because of the long term repeatability requirements a digital controller is used to close the loop. As the equipment is for use by unskilled operators the complete test sequence is fully automated and error checking routines are built into the controller to ensure that each test is within specification.

10.2 A High-Rate Impact Tester

Impact resistance is a key performance requirement for plastic products. Previous test methods suffered from lack of control of strain rate of the test specimen. This problem has been solved by the use of electrohydraulic control.

The machine must operate over a wide speed range of 0.75 to 750 m/min. This high speed requirement associated with the relatively high force level of around 2,000 to 4,000 N represents a peak power level of the order of 50 kW. The actuator used to achieve this is an assembly of a cylinder and two high performance electrohydraulic valves, see figure 9.12.

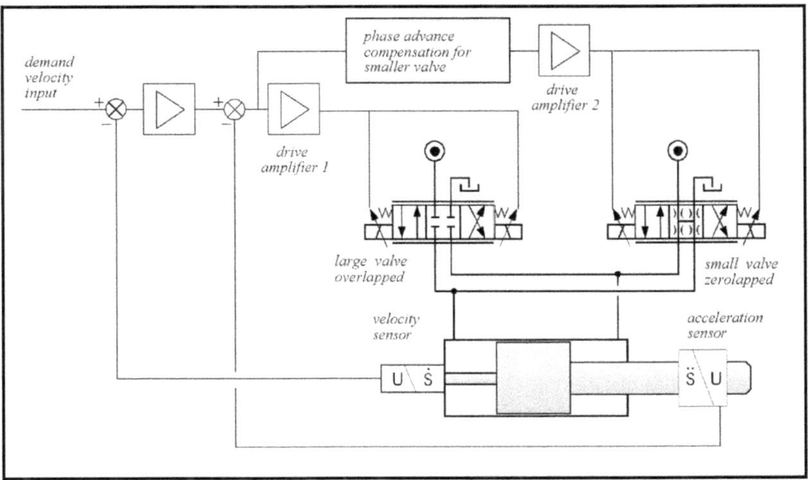

Figure 9.12: High Rate Impact Tester

The actuator consists of a rugged aluminium cylinder with a chrome plated steel piston. For a constant accelerating force, the mass of the piston and the impacting mechanism determine the acceleration distance. The design uses a hollow piston rod with an integral piston head. The actuator is equipped with snubbing capability to stop a runaway piston without damage. The primary feedback is from an inductive velocity transducer. The transducer is totally enclosed and mounted co-axially within the actuator eliminating backlash and high vibration sensitivity of an offset transducer location.

An accelerometer is located on the piston rod which, through its feedback, provides damping of the actuator resonance since this resonance is the factor which limits the achievable loop gain.

The wide speed range is very difficult to meet with one valve. Consequently, two valves, a two-stage and three-stage valve, are connected hydraulically in parallel. Smooth operation at low speeds requires that the three-stage valve be effectively isolated. An appropriately selected overlap on the three-stage valve allows only the small two-stage valve to control flow to the piston at low speeds. The three-stage valve starts supplementing the small valve once the latter reaches its full flow capacity. A fast responding valve is required to keep the acceleration distance short and to have a minimum phase lag at the actuator resonant frequency to provide the capability of actively damping this resonance.

The primary objective of the system is to control the speed of the piston before, during and after the impact. Comparable performance over the wide speed range is achieved by designing the small amplitude response of the two electrohydraulic valves to be similar. A high response pilot valve and lead compensation in the third stage spool position loop are used to achieve this.

10.3 Precise Speed Control of a Centrifuge

The Delft Institute for Soil Mechanics has the world's largest centrifuge used for the investigation of ground conditions by imposing loads on miniaturised models. The centrifuge has a 6 m long rotor having an overall weight of 5,500 kg which is rotated up to speeds of 210 rev/min by two hydraulic motors using a total power of 400 kW. This produces a circumferential speed of 500 km/hour and an acceleration of 300 g.

Accurate speed control is a major requirement. This is held to ±0.05% of actual speed within the speed range of 10 to 100% of maximum. In addition to the accuracy of speed control the system must be maintain constancy and repeatability of the drive speed over long periods. These aims could only achieved by using a digital encoder on the centrifuge drive shaft to measure centrifuge position and a stable quartz controlled oscillator to generate digitally a ramp signal of position demand. The result from this high accuracy digital comparison is converted to an analogue signal which forms the input to an analogue speed control loop using a tachogenerator on the centrifuge shaft to supply the feedback signal. The error from this loop is used to drive the motor swash control loops which are also analogue.

Two power matched hydraulic systems drive the rotor arm via 6:1 ratio reduction gearboxes, see figure 9.13. The two motors are controlled by separate electrohydraulic valves and closed-loop control systems to control accurately the swash angle. The

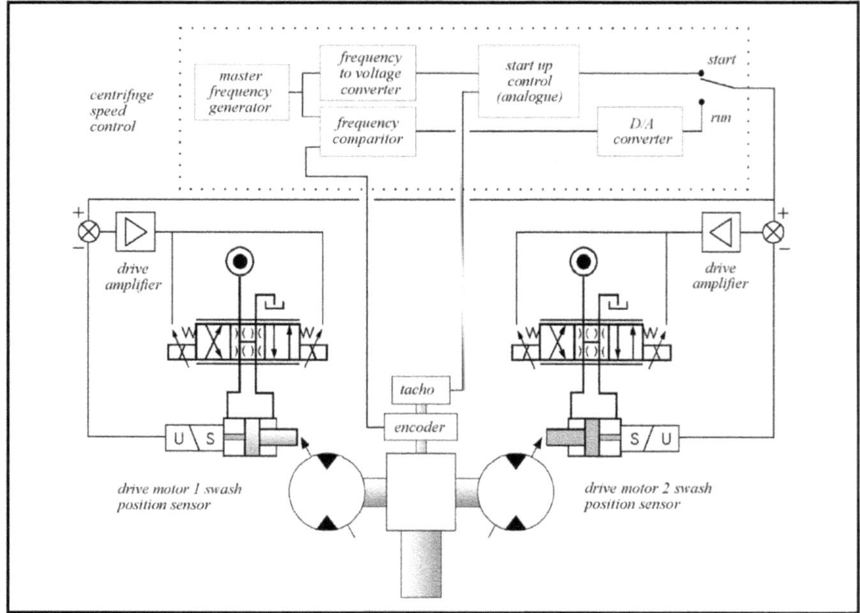

Figure 9.13: High Precision Speed Control Circuit

systems are linked so that the swash angles of the two motors can be closely synchronised at all times.

Two small hydraulic pumps are directly coupled to each motor shaft and these supply all system pilot flow requirements. This arrangement was chosen on safety grounds. If the electrical supply should fail, the pilot supply to the electrohydraulic valves is maintained as they obtain their power from the kinetic energy of the centrifuge whilst it continues to rotate.

The acceleration and deceleration phases are performed using analogue speed control mode. The command value for the speed is passed to the control via a digital ramp. The acceleration and deceleration times can be present within the range 3 to 33 rev/min^2. Once the centrifuge has reached a stable condition, the analogue speed control loop becomes subsidiary to the digital position loop.

In the deceleration mode these motors act as pumps, feeding power back via the power supply pumps into the electrical power supply.

11 REFERENCES

ISO 4413 Hydraulic fluid power - General rules relating to systems

BS EN 982 Safety of machinery - Safety requirements for fluid power systems and their components - Hydraulics

REFERENCE SECTION

GLOSSARY OF TERMS
RECOMMENDED SYMBOLS AND UNITS
READING LIST

GLOSSARY OF TERMS

Acceleration feedback – A form of feedback compensation where a signal proportional to acceleration may be used to give improved damping.

Amplitude ratio – Ratio of output amplitude to input amplitude at a particular sinewave frequency. It may be applied to components or systems and is often expressed using a scale in decibels (dB).

Analogue – A signal which may vary continuously with time, usually used to distinguish from a digital signal.

Bandwidth – A frequency range over which a specified response can be obtained. It is usually specified relative to a 3dB change in amplitude (70% amplitude ratio) or for a 90° change in phase. It is often used as a guide to the maximum frequency of operation.

Buffer amplifier – An electrical amplifier which separates two parts of a circuit frequently without changing signal level i.e. the output voltage is equal to input voltage. Usually used to prevent excessive loading of the signal source.

Charge amplifier – An electrical amplifier which produces a voltage output proportional to charge input. Mainly used with Piezoelectric devices.

Closed-loop – Operation of a system with feedback connected.

Compensation – A network (usually electrical) which corrects an undesirable characteristic. For closed-loop systems it can be used to improve response in either dynamic or steady state operation. It may be either in the forward path or in the feedback path.

Continuous control valve – Valve that controls the flow of fluid energy, in a system, in a continuous way, in response to a continuously variable input signal. Note: This encompasses all types of servo-valve and proportional control valves.

Continuous directional control valve – A continuous control valve with reversal of flow direction according to the polarity of the input signal. This encompasses all types of single or multi stage servo-valve and proportional directional control valves.

Continuous flow control valve – A continuous control valve producing a controlled flow rate at one control port and in one direction only.

Continuous pressure control valve – A continuous control valve producing a controlled pressure at the control port.

Control pressure – The pressure existing at either control port, which is normally variable between supply pressure and return pressure.

Current drive amplifier – An amplifier which gives an output as a current level which is directly proportional to the input (signal) level.

Damping ratio – The ratio of actual damping in a system to the critical damping. Strictly should be only applied to systems described by second order differential equations but very often applied more widely where second order behaviour is dominant.

Deadband compensation – The automatically generated signal that is added to the input signal to compensate hydraulic overlap of the valve.

Decibel (Db) – A logarithmic scale of two signals, strictly describing the power ratio. Often used to ratio two voltage signals where dB level is given by 20 log10 (V out/ V in).

Demodulator – A circuit used in position and pressure sensors to extract an output signal from an amplitude modulated fixed frequency wave form.

Derivative control – A form of compensation where the derivative of a signal is used to improve performance.

Difference amplifier – An amplifier which gives an output proportional to the difference between two input signals. Often called differential amplif-ier.

Dither – A periodic signal that is superimposed on a drive signal to minimise hysteresis and friction effects, usually at a low amplitude and higher frequency than the system bandwidth.

Dynamic response – The ability of a system to follow a rapidly changing input signal, measured either by a step response or frequency response.

Electrically modulated hydraulic control valve – See 'Continuous control valve'.

Electrohydraulic valve – A type of continuous directional control valve, continuous flow control valve or continuous pressure control valve having an electrical input.

Eurocard – A range of standard sized electronic circuit cards for mounting in equipment racks.

Feed forward – A type of compensation where the demand is passed directly into the forward path avoiding the summing junction, sometimes a transient form of compensation.

Force motor – A type of electro-mechanical transducer having linear motion in which the direction of force changes with the direction of current.

Forward path gain – A gain in the forward path of a control loop, could be amplifier gain or more strictly, the net gain from error to output.

Four-port flow control valve – Multi-orifice variable flow control valve with supply, return and two control ports arranged so that the valve action in one direction throttles supply to control port A (2) and throttles control port B (4) to return. Reversed valve action throttles supply to control port B and throttles control port A to return.

Frequency response – The relationship between an output signal and an input signal as the input is varied as a sinewave over a range of frequencies. Expressed in terms of the amplitude ratio of output to input and as a phase shift between output and input. May be normalised by using a specified low frequency response as a datum.

Gain – Relationship between the output and input expressed as a ratio. Graphically, it can be considered as the slope of the output to input characteristic. For non-linear characteristic, it has to be stated for which region the gain applies.

Hydraulic amplifier – A fluid device which acts as an amplifier. Hydraulic amplifiers may utilize sliding spools, flapper-nozzles, jet pipes, etc.

Hydraulic null – The condition where the valve supplies zero control flow. Does not apply to pressure control valves.

Hysteresis – Difference in controlled parameters, at the same control setting, when adjusting the quantity upwards and then downwards, or vice versa.

Inlet pressure – Pressure at the inlet port of a component, piping or system.

Integral control – Type of forward path compensator where the error signal is integrated to reduce steady state errors.

Intrinsic safety – A requirement for electrical equipment to operate safely in potentially explosive atmospheres.

Lap – In a sliding spool valve, the relative axial position relationship between the fixed and movable flow metering edges with the spool at null.

Overlap – A condition where the fixed and movable flow metering edges do not coincide with the spool at null in such a way that a relative displacement between the metering edges must occur before an effective flow path is created. The effect of overlap on a valve having rectangular orifices is shown in figure 2.3.

Underlap – A condition where the fixed and movable flow metering edges do not coincide with the spool at null in such a way that a flow path exists across two or more metering edges with the spool at null.

Zero lap – A condition where the fixed and movable flow metering edges coincide with the spool at null. In a valve having rectangular orifices this condition would give rise to substantially constant flow gain over the null and operational regions. See figure 2.1.

Lead compensator – Type of forward path compensator producing phase lead to improve the dynamic response characteristics.

Load cell – Sensor to produce a signal proportional to applied force.

Loop gain – The total gain around the signal path of a closed-loop system.

Low pass filter – An electrical circuit which removes high frequency components of a signal and has minimal affect on the low frequency components of a signal.

Maximum working pressure – Highest pressure at which a system, or sub-system is intended to operate in steady-state conditions.

Natural frequency – Frequency at which a system will vibrate in free oscillation dependent principally on mass and stiffness.

Null bias – The input signal required to bring the valve to hydraulic null. (See "Hydraulic Null").
Null pressure – The pressure existing at both control ports at hydraulic null.
Null region – The region about null where the flow gain is affected by parameters such as lap and internal leakage.
Null shift – The change in null bias required as a result of a change in operating conditions or environment or long term effects with tolerance to the input signal.

Open-loop – A system which operates without feedback. Also applied to a system or system transfer function which contains all the components for closed-loop operation but without the connection at the summing junction.

Phase lag – A phase shift which means that for a sinusoidal signal the output occurs after the input.
Phase lead – A phase shift which means that for a sinusoidal signal the output occurs before the input.
PID – Proportional plus Integral plus Derivative control action. A common form of compensation to improve both steady state accuracy with the integral effect and the dynamic response with the derivative effect.
Pilot flow – The fluid flow used to control the mainstage of a two-stage valve.
Pressure gain – The change in differential pressure between the control ports per unit input signal with zero control flow (control ports blocked). Pressure gain is specified as the average slope of the curve of load pressure drop versus input signal in the region between ± 40% of supply pressure.
Proportional control – Closed-loop control with a constant gain term in the forward path.
Proportional solenoid – A solenoid constructed to give a force output proportional to input current and largely independent of stroke.
Proportional control valve – A type of continuous directional control valve.
Pulse width modulation – A PWM signal is a switch signal where the ratio of ON time to OFF time is varied to give a mean value proportional to the input.

Quiescent current – The current taken by a valve to maintain the null position.

Ramp input – An input signal increasing or decreasing linearly with time.
Rated current – The current to give the specified output from a valve where this output may be flow or pressure.
Rated flow – The control flow corresponding to the maximum value of the input signal and nominal valve pressure drop. If no pressure drop is specified, the rated flow is equivalent to maximum flow.
Rated pressure – Pressure confirmed through testing, at which a component, or piping, is designed to operate for a number of repetitions sufficient to assure adequate service life – i.e. the figure normally quoted in manufacturers' catalogues.
Rated valve pressure drop – The valve pressure drop at which rated flow is measured, typically selected as 35 bar per land (servo-valve) or 5 bar per land (proportional valves).

Resolution – The change of input signal required to produce a change of the output quantity. Starting from a rest point the input is changed in the same direction as the one when approaching the rest point.

Response time – The time taken for the output to reach a specified percentage for the first occasion of its rated value if the valve is subjected to a step input.

Return pressure (or Back pressure) – The pressure at the return port of the valve.

Servo-valve – A type of continuous directional control valve

Stability margin – The concept in control engineering which identifies how near a closed-loop system is to a region of operation where it will become unstable even without a demand signal.

Step response – The response of the output quantity in the time domain as a result of a step input signal.

Summing amplifier – An amplifier circuit which will add together two or more signals.

Supply pressure – Pressure generated by a power supply.

Three-port flow control valve – A multi-orifice variable flow control valve with supply, return and one control port arranged so that valve action in one direction throttles supply to control port and reversed valve action throttles the control port to return.

Threshold – The change of input signal required to produce a change of the output quantity. Starting from a rest point the input is changed in the opposite direction to the one when approaching the rest point.

Torque motor – A type of electro-mechanical transducer having rotary motion in which the direction of motion changes with the direction of current.

Transfer function – A mathematical function expressing the relationship between a system input and its output.

Transient – A condition which decays with time and does not persist. In a transient response the output to input relationship is changing with time.

Two-port flow control valve – A variable flow control valve with a single throttling orifice between the two-ports.

Valve drive amplifier – An amplifier which gives an output appropriate to a valve type, usually this would be a current output and may be PWM operated.

Valve pressure drop – For flow control valves: The valve pressure drop equals the supply pressure minus the return pressure minus the pressure required to overcome the load.

Velocity feedback – A form of feedback compensation where the derivative of an output position may be used as an additional feedback loop.

RECOMMENDED SYMBOLS AND UNITS

Parameter	Symbol	Unit
Coil Impedance	Z	Ohm
Coil Resistance	R	Ohm
Dither	I_c	-
Input Current	I	mA
Quiescent Current	I_q	mA
Rated Current	I_n	mA
Electrical Control Power	P_n	Watt
Electrical Quiescent Power	P_q	Watt
Total Electrical Power	P_t	Watt
Control Flow	q	Litre/min
Flow Gain	K_v(dq/dI or dq/dy)	Litre/min/mA or Litre/min/mm
Hysteresis	η	-
Internal Leakage	q_l	Litre/min
Lap	x_l (+ for underlap - for overlap)	mm
Valve Travel	y	mm
Valve Opening	x	mm
Supply Pressure	P_s	bar
Tank Pressure	Pt	bar
Control Pressure	Pa or Pb	bar
Valve Pressure Drop	ΔP_v	bar
Pressure Gain	S_v (dp_l/dI or dp_l/dy)	bar/mA or bar/mm
Resolution	ηr	-
Threshold	ηt	-
Amplitude Ratio	L_m	dB
Phase Lag	Ø	degrees

Note: 1 bar = $10^5 N/m^2$

READING LIST

ANALYSIS, SYNTHESIS & DESIGN OF HYDRAULIC SERVOSYSTEMS
T J Viersma, Elsevier Scientific
ISBN 0444418695

AUTOMATIC CONTROL ENGINEERING
F H Raven, McGraw-Hill, Kogakusha
ISBN 0070512337

CONTROL OF FLUID POWER (ANALYSIS AND DESIGN)
D McCloy & H R Martin, Ellis Horwood, John Wiley
ISBN 0853122482 (out of print)

CONTROL SYSTEM TECHNOLOGY
C H Chesmond, Edward Arnold
ISBN 00713135085

CONTROLLING ELECTROHYDRAULIC SYSTEMS
W R Anderson, Marcel Dekker
ISBN 0824778251

DESIGN OF ELECTROHYDRAULIC SERVO SYSTEMS FOR INDUSTRIAL MOTION CONTROL
J Johnson, Penton Education Division

ELECTRONICALLY CONTROLLED PROPORTIONAL VALVES
M J Tonyan, Marcel Dekker
ISBN 0-8247-7431-0

FEEDBACK CONTROL OF DYNAMIC SYSTEMS
G F Franklin, J D Powell, A Emami-Naeini, Addison-Wesley
ISBN 0-201-11540-9

FLUID POWER DESIGN HAND-BOOK
F Yeaple, Marcel Dekker
ISBN 0824771966

FLUID POWER SYSTEMS
J Watton, Prentice Hall
ISBN 0-13-323197-6

GUIDELINES TO CONTAMINATION CONTROL IN HYDRAULIC FLUID POWER SYSTEMS
BFPA/P5
British Fluid Power Association

HYDRAULIC CONTROL SYSTEMS
H E Merritt, John Wiley
ISBN 0471596175

HYDRAULIC & ELECTRO-HYDRAULIC CONTROL SYSTEMS
R B Walters, Flotron Ltd
ISBN 0-7923-6537-2

HYDRAULIC SERVOSYSTEMS
Guillon Butterworth (out of print)

HYDRAULICS - THEORY AND APPLICATIONS
Bosch Rexroth Ltd

INDUSTRIAL HYDRAULICS MANUAL
Eaton Hydraulics Ltd

MANAGEMENT OF INDUSTRIAL MAINTENANCE
A Kelly & M J Harris
Butterworth, ISBN 040801377x

MODELLING & SIMULATION SOFTWARE FOR FLUID POWER TECHNOLOGY
D Harrison, R B Walters, Flotron Ltd
International Journal of Fluid Power, Volume 4 Number 3
ISSN 1439-9776

PRINCIPLES OF AUTOMATIC CONTROL
M Healey, Hodder & Stoughton
ISBN 034176717

PRINCLIPES OF HYDRAULIC SYSTEM DESIGN
Peter J Chapple
ISBN 1 901892 15 8
BFPA/P95

RELIABILITY AND MAINTAINABILITY IN PERSPECTIVE
D J Lewis, Macmillan Press Ltd
ISBN 033331049

SAE HANDBOOK
Published annually, Society of Automotive Engineers

SYSTEM MODELLING AND CONTROL
J Schwarzenbach & K F Gill, Edward Arnold,
ISBN 0713135182

THE HYDRAULIC TRAINER
A Schmidt, G L Rexroth GmbH

TRIBOLOGY HANDBOOK
Ed: M J Neale
Butterworths

USING INDUSTRIAL HYDRAULICS
T C Frankenfield, Penton Education Division

ISO 4413 Hydraulic fluid power - General rules relating to systems

BS EN 982 Safety of Machinery - Safety requirements for fluid power systems and their components - Hydraulics

INDEX

Index

Symbols

A

B

F

J

K

L

M

Q

R

Y

Z

www.ingramcontent.com/pod-product-compliance
Ingram Content Group UK Ltd.
Pitfield, Milton Keynes, MK11 3LW, UK
UKHW021312070726
13610UKWH00001B/2